W9-CRB-818

Food Taste Chemistry

Food Taste Chemistry

James C. Boudreau, EDITOR
University of Texas—Houston

Based on a symposium
sponsored by the Division of
Agricultural and Food Chemistry
at the ACS/CSJ Chemical Congress,
Honolulu, Hawaii,
April 2–6, 1979.

ACS SYMPOSIUM SERIES **115**

AMERICAN CHEMICAL SOCIETY
WASHINGTON, D. C. 1979

Library of Congress CIP Data

Food taste chemistry.
(ACS symposium series; 115 ISSN 0097-6156)

Includes bibliographies and index.

1. Food—Analysis—Congresses. 2. Flavor—Congresses. 3. Taste—Congresses.
I. Boudreau, James C., 1936- . II. American Chemical Society. Division of Agricultural and Food Chemistry. III. Series: American Chemical Society. ACS symposium series; 115.

TX511.F685 612'.87 79-26461
ISBN 0-8412-0526-4 ASCMC 8 115 1–262 1979

PRINTED IN THE UNITED STATES OF AMERICA

ACS Symposium Series

M. Joan Comstock, *Series Editor*

FOREWORD

The ACS Symposium Series was founded in 1974 to provide a medium for publishing symposia quickly in book form. The format of the Series parallels that of the continuing Advances in Chemistry Series except that in order to save time the papers are not typeset but are reproduced as they are submitted by the authors in camera-ready form. Papers are reviewed under the supervision of the Editors with the assistance of the Series Advisory Board and are selected to maintain the integrity of the symposia; however, verbatim reproductions of previously published papers are not accepted. Both reviews and reports of research are acceptable since symposia may embrace both types of presentation.

CONTENTS

PREFACE

The symposium on which this book is based was one of the few symposia, ever, totally devoted to the chemistry of food tastes. Most previous symposia have dealt with food flavors, where flavor is considered to consist mostly of odorous sensations.

Taste has long been considered to consist of only four sensations that contribute little to most food flavors. These four feeble sensations were linked to a simplistic taste chemistry that had little relevance to modern chemistry. These conceptions, often repeated, not only totally ignore the major role taste plays in food selection and the control of ingestion, but also are not followed in practice by much of the flavor industry. Thus you will often discover upon reading the literature that the "odors" of a food were best, or perhaps only, realized when food was in the mouth. Many flavor chemists have found that in order to adequately define a food flavor, tastes other than the four basics must be postulated. The types of taste active compounds in foods encompass much of natural product chemistry. Many of the compounds presently identified as odors are strongly taste active.

To help lay a new groundwork for the study of the tastes of foods, the speakers at the symposium presented papers on a variety of topics related to food taste chemistry. The problems in taste are of great complexity, involving biological as well as chemical variables. For a taste chemist, the types of sensations elicited and their measurement are as important as the nature of the compounds eliciting them. Various aspects of these problems are treated in detail in the papers in this volume.

The findings presented at this symposium have relevance far outside the narrow area of flavor chemistry. The taste measurements of the chemical properties of nutrient solutions have applications that range from physical organic chemistry to human nutrition.

Special thanks are due to the Japanese cochairman M. Namiki and the members of the Agricultural and Food Chemistry Division of ACS, especially G. Charalambous, R. Teranishi, and C. Mussinan for organizing and scheduling the symposium. I thank J. Oravec, Ng. Hoang, and the ACS Books Department for assistance in preparing this volume for publication.

University of Texas—Houston
Houston, TX 77025
September 13, 1979

JAMES C. BOUDREAU

1

Taste and the Taste of Foods

JAMES C. BOUDREAU, JOSEPH ORAVEC, NGA KIEU HOANG, and THOMAS D. WHITE

Sensory Sciences Center, Graduate School of Biomedical Sciences, University of Texas at Houston, Houston, TX 77030

In this paper the term taste will refer to all the chemical sensory systems of the oral cavity and their sensations. These sensory systems are intimately involved in the selection of food items and in the regulation of food intake. As we shall see, there are a variety of different taste systems attuned to different chemical aspects of food. These taste systems perform an exact and elaborate analysis of the chemical constituents in the food we eat.

The structure and function of these taste systems will be discussed in the context of a natural nutritional ecosystem, i.e. one in which man is not a disruptive element. Human taste systems are assumed to have developed to function in this natural system and to have changed little as a result of the cultural dietary changes that have occurred in the last 10- 20,000 years.

The natural nutritional ecosystem of man is assumed to be one in which both plant and animal foods are eaten (Figure 1) and they are eaten raw. In a natural nutritional ecosystem, taste serves a primary role in regulating the flow of compounds (1, 2). Certain things are to be eaten by us. Other things by others. There exists wide variation in the tastes of natural foods. Thus Cott has shown that certain birds and their eggs are both conspicuous and ill tasting (3, 4). The types of foods we consume now represent a selection from the vast array of items naturally available during our evolutionary development. The chicken egg for instance, represents a selection of one of the best tasting eggs naturally available (Figure 2).

We consume and transform plant and animal substances to promote certain physiological activities (the probable role of taste in mammalian sexual behavior is not considered here). Primary among these physiological functions is the replacement of body compounds and the supply of compounds for metabolic energy systems. Thus taste serves to regulate the consumption of needed compounds. Almost without exception, natural things that taste good are good for you and foods that are needed taste good even though your stomach is full. Toxic compounds almost

0-8412-0526-4/79/47-115-001$08.00/0

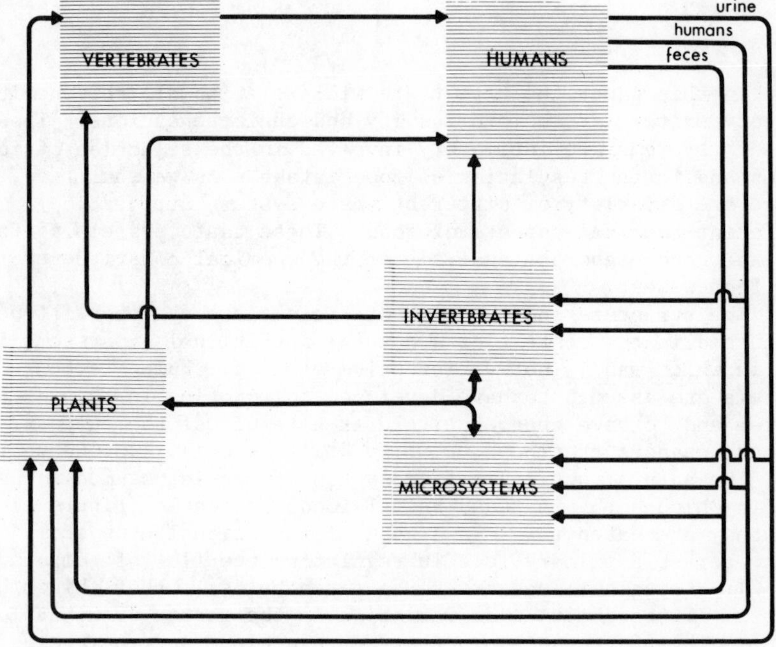

Figure 1. Flow diagram of the natural nutritional ecosystem of the human (simplified)

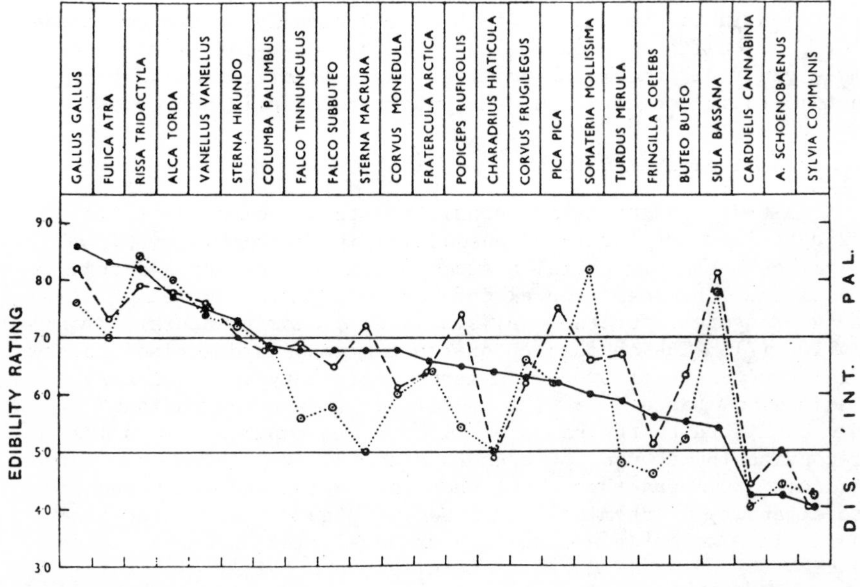

Figure 2. Preferences of man (●—●), rat (○ - - - ○), and hedgehog (○ · · · ○) for some eggs of different species of birds (4). The species Gallus gallus *is the chicken.*

invariably have noxious tastes. One exception (omitting marine substances) is the poisonous Amanita phaloides mushroom, a fungus that tastes good but will kill you. Not only do toxic foods have noxious tastes, but the thresholds for many toxic substances are extremely low. Another possible function of taste is the ingestion of compounds for the regulation of body temperature. Although there seems to exist little hard data on this matter, many human cultures classify foods into those that warm the body and those that cool it (5, 6). Taste may also function in the selection of pharmacologically active compounds for good health or good feeling. Things that taste good often make you feel good. In addition, many flavor compounds have antimicrobial actions and other pharmacological properties.

Anatomy of Taste Systems

A taste system can be considered to be composed of a receptor element for the transduction of chemical signals, a peripheral sensory neural system for the collection and transmission of chemical neural information, and a complex central nervous system for the analysis of this sensory neural information (7). The chemoreceptors that have been described in the oral cavity are of two basic morphological types: free nerve endings and taste buds. The so-called "free nerve endings" are distinguished on the basis of light microscopy as possessing no recognizable receptor or encapsulated ending. These free nerve endings are found throughout the oral cavity and are responsive to a variety of chemical compounds. A taste bud, on the other hand, is a receptor neural complex consisting of nerve fibers and 20-50 specialized cells organized in a fairly elaborate manner (Figure 3). The elongated taste bud cells are grouped together with one end forming the floor of the taste pit which opens up, through the taste pore, to the oral fluids. The taste bud cells project into the taste pore with either microvilli or an elongated bulb. The taste bud cells have been classified morphologically into three or more distinct types (8-12).

Taste buds, unlike free nerve endings, are not distributed throughout the oral cavity but rather are on the dorsum of the tongue, the soft palate, pharynx, epiglottis, larynx and upper third of the esophagus (Figure 3). On the tongue, taste buds are localized on protuberances known as papillae. The taste buds on the front two thirds of the tongue are located on the dorsal surface of the small fungiform papillae. At the rear of the tongue the taste buds are located in the foliate papillae and the vallate papillae. The posteriorly located chemosensory complexes contain large numbers of taste buds together with specialized secretory glands.

The peripheral sensory neurons that supply the chemoreceptors in the oral cavity reside in four distinct cranial ganglia (Figure 4). The trigeminal ganglion contains the sensory

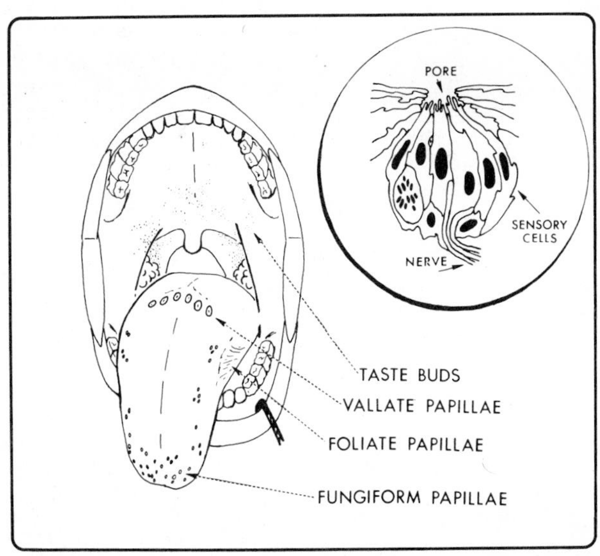

Figure 3. Location of some oral chemosensory receptor systems. Taste buds (schematic upper right) are found on specialized papillae on the tongue and scattered on the palate and posterior oral structures. Free nerve endings are found on all oral surfaces (94).

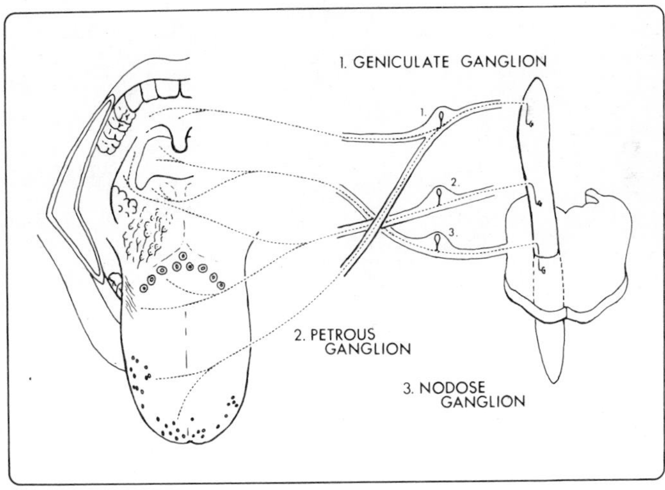

Figure 4. Peripheral sensory ganglia that supply nerve endings to taste buds in the mammalian oral cavity. Trigeminal ganglion, which supplies free nerve endings to all oral surfaces, not shown.

neurons providing free nerve endings to all parts of the oral cavity. The other three sensory ganglia innervate the taste buds, with each ganglion innervating buds on distinct locations. The taste buds on the fungiform papillae and the anterior soft palate are innervated by sensory neurons in the geniculate ganglion of the facial nerve. The taste buds on the foliate papillae, the circumvallate papillae, the posterior palate, the tonsils and the fauces are innervated by cells in the petrous ganglion of the glossopharyngeal nerve. Taste buds on the epiglottis, the larynx and the upper third of the esophagus are innervated by neurons in the nodose ganglion of the vagus nerve. Physiological and psychophysical studies on the functional properties of these different nerves and ganglia indicate that the chemosensory systems in the different ganglia are selectively responsive to different chemical aspects of foods.

Neurophysiology of Taste Systems

In examining the function of taste systems, various physiological measures are available to the investigator. Although, theoretically the neurophysiological responses of either the receptors or from any of the neurons in the sensorineural chain may be utilized, in practice, the most exact procedure is to measure the pulse trains being transmitted from the periphery to the central nervous system (Figure 5). Receptor potentials are subject to several sources of error, at least as regards quantitative measures of neural responses. For the precise study of neural responses to chemical stimulation the neural pulse is the measure of choice in sensory neurophysiology. These pulses are preferably measured from peripheral sensory neurons since neural interaction is minimized. Pulses may be measured from either the peripheral fibers of the sensory ganglion cells or from the somas. One advantage in recording pulses from the cells themselves is that small fiber systems are sampled, while there is a strong bias toward large fiber potentials in fiber recordings. The only ganglion cell system that has been examined in any detail is the geniculate ganglion system which innervates receptors on the fungiform papillae. The properties of these neurons will be reviewed for the cat, dog, and goat.

Typically, a geniculate ganglion cell innervates receptors on more than one fungiform papilla (Figure 5). The number of fungiform papillae innervated by a single neuron ranges from one to as many as twelve. Within a taste bud, a nerve fiber will contact many receptor cells. Almost all geniculate ganglion neurons exhibit pulse activity in the absence of experimenter designed stimulation (Figure 6). This "spontaneous activity" is usually of a complex irregular type that is characteristic of chemical sensory systems. Pulses are often emitted in bursts with fixed interspike intervals, with the burst interval

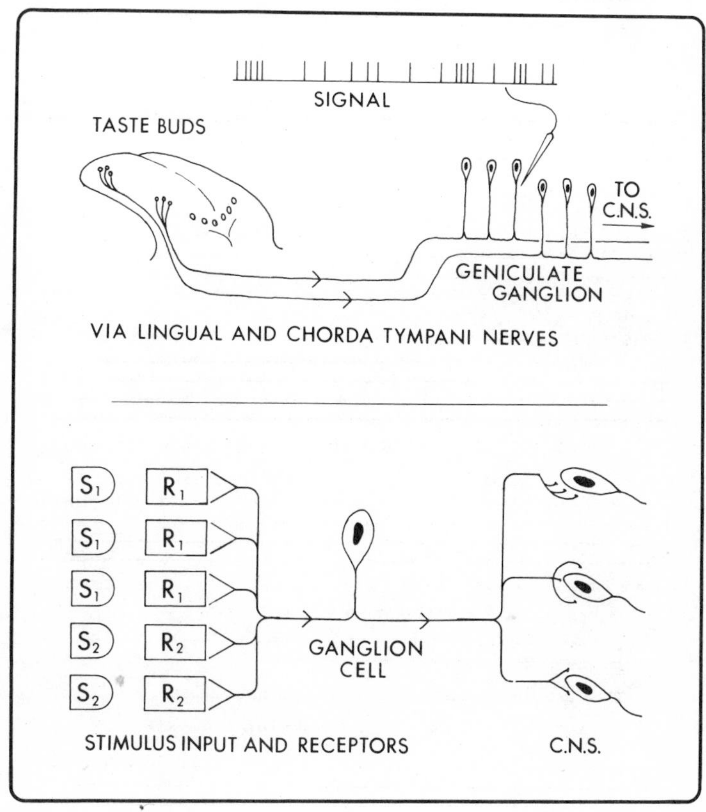

Figure 5. (lower) *Diagram of the peripheral and central connections of a sensory ganglion cell innervating the taste buds of the tongue.* (upper) *Illustration of the connections of sensory ganglion cells and the pulse signals used to encode sensory information.*

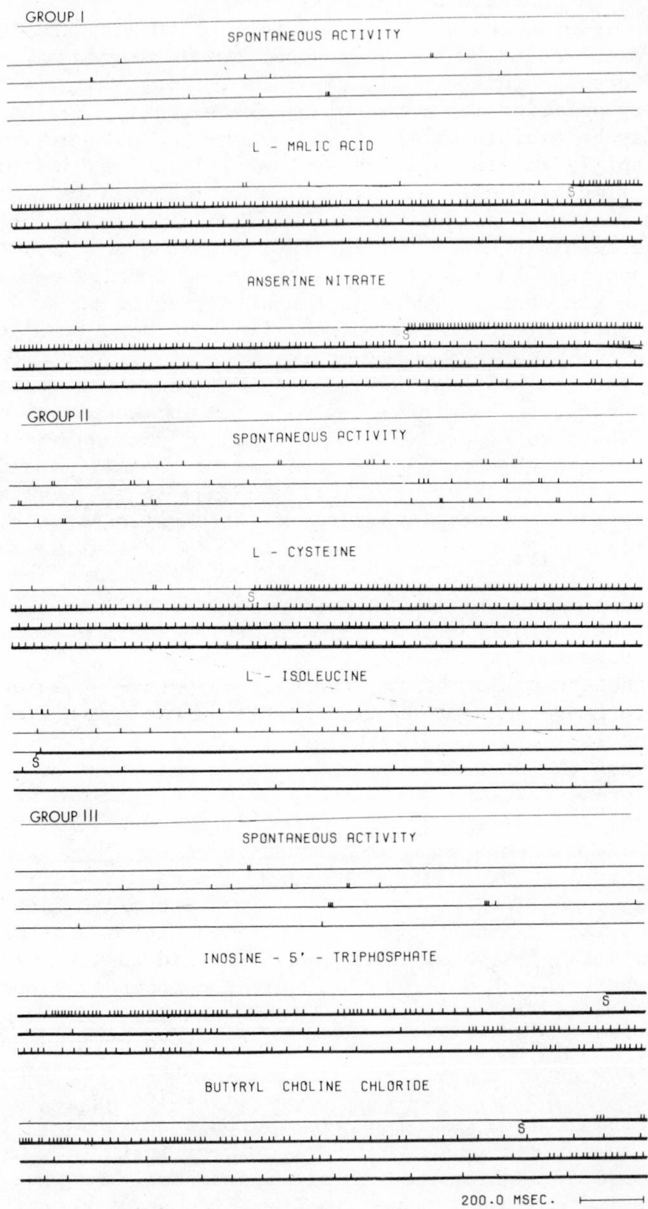

Figure 6. Spontaneous and evoked spike activity recorded from taste neurons of the geniculate ganglion of the cat. The classification of the three different sensory neurons is indicated by Groups I, II, and III.

decreasing as function of interval order.

This spontaneous activity may be inhibited by the application of a chemical solution to the tongue, or the neuron may be excited by a different solution. Mixing an inhibitory compound into an excitatory solution may result in an inhibitory solution. The neuron may be excited or inhibited by chemical stimulation of a single papilla in its papillae system (13), and excitation of two papillae simultaneously may result in an increase in discharge, with the increase being usually less than algebraic. The discharge resulting from the excitation of the neuron by stimulation of one papilla may be inhibited by the simultaneous stimulation of another papilla with an inhibitory solution.

Neurons of the geniculate ganglion have been found to be of more than one type when examined in terms of various physiological measures such as spontaneous activity rate and type, latency of spike discharge to electrical stimulation (a measure of conduction velocity and thus fiber size), and types of compounds activating. Neurons in both the cat and the dog can be classified into at least three different groups (10). Neurons in the goat have also been tenatively divided into three different groups, although only one of these groups is comparable to those in the two carnivores. The neural groups in the cat, goat, and dog tend to preferentially innervate fungiform papillae on somewhat different parts of the tongue (Figure 7), although there is extensive overlap, especially in the dog.

The determination of the types of compounds stimulating geniculate ganglion neurons constitutes an extensive field of continuing investigation. Not surprisingly, it has been found that many of the neurons are sensitive to solutions of foods commonly present in the animal's environment. Thus a goat neuron may respond to a carrot or herb solution and a cat to chicken or liver. Cats and dogs have been found to be highly responsive to many of the compounds found in meats, such as amino acids, the dipeptides, anserine and carnosine, and nucleotides. The goat has been less well investigated, but seems highly responsive to salts and alkaloids. Especially prominent in the stimulation of the carnivore are nitrogen and sulfur compounds, especially five and six member ring heterocycles. The different neural groups tend to be differentially responsive to chemical stimuli, illustrating their selectivity in the measurement of food compounds (Figure 8). Some of the similarities and differences among the geniculate ganglion neural groups of the cat, dog, and goat are summarized in Table I. As evident in this table, the dog geniculate ganglion systems are quite similar to the cat. One major distinction is that the dog amino acid sensitive neurons (class A units), although highly similar to cat group II units, are also responsive to sugar as well as the most stimulating amino acids and di- and tri- phosphate nucleotide salts. The goat neural groups on the other hand seem quite distinct from carnivoral taste systems with only the acid responsive group

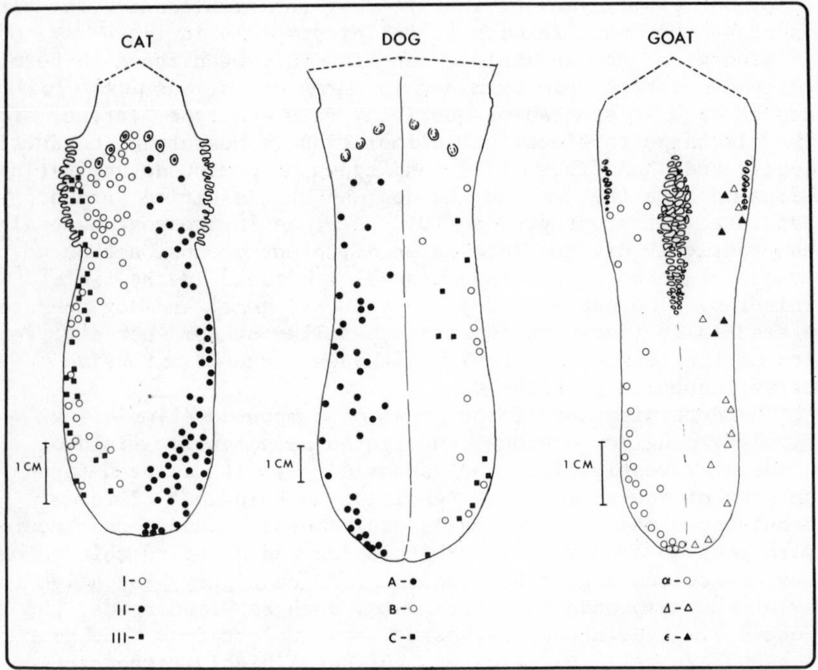

Figure 7. Peripheral innervation of fungiform chemoreceptors by neurons of the geniculate ganglion of three different species. In each species the neurons have been separated into three distinct neural groups (see Table I for comparison of chemical stimuli for the different neural groups).

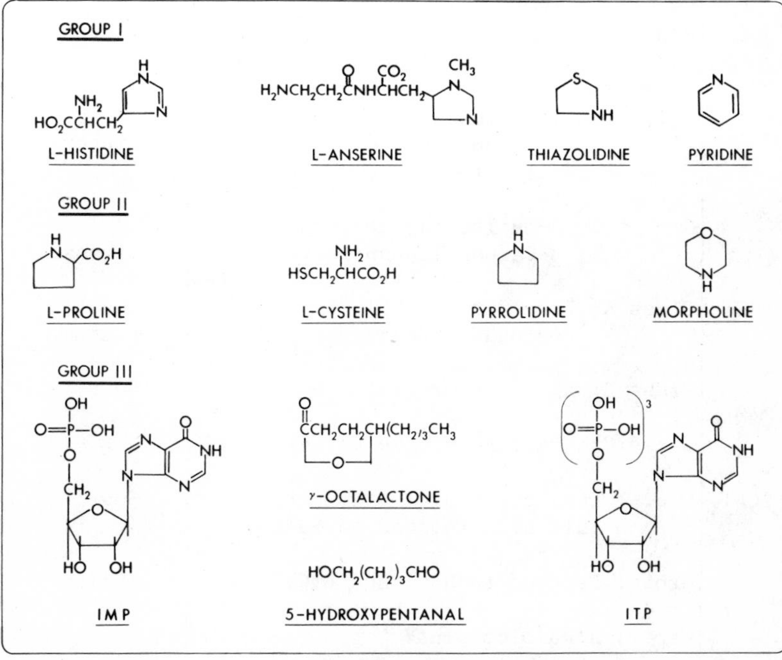

Figure 8. Chemical formulas for some of the most active stimuli for the cat geniculate ganglion neural groups

Table I: Summary of Neurophysiological Investigations on Mammalian Geniculate Ganglion Taste Systems.

Species	Cell Groups and Stimuli	Human Sensations
CAT	Grp. I: Organic Acids and Histidine Compounds. Also to Alkaloids	Sour
	Grp. II: Amino acid responsive, di and triphosphate nucleotides, NaCl potentiated	Sweet[1] Bitter[1]
	Grp. III: A-Nucleotide Responsive B-Carbonyl Responsive	- (Pleasant?)
DOG	Class A: Like Cat grp. II, except respond to sugars	Sweet[1] Bitter[1]
	Class B: Like Cat Group I	Sour
	Class C: Partially like Cat Grp. III	-
GOAT	Grp. : Like cat grp. I & dog grp. B, only also respond to Salts	Sour
	Grp. : Respond to NaCl and LiCl	(Salty?)
	Grp. : Alkaloids plus?	-

common with the other two species. No taste system highly responsive to amino acids has been found in the goat; and, unlike the carnovore, large numbers of neurons, including the acid responsive system, are discharged by NaCl.

Although the other major tongue taste system - that represented by the petrous ganglion of the glossopharyngeal nerve-has been examined in mammals by several investigators, they have not utilized a sufficient number of chemical compounds for us to determine their possible role in the measurement of food compounds. One study (15) in the frog glossopharyngeal nerve (not directly comparable to mammalian) found that many compounds assumed to be functioning as odors were strong stimuli. Thus active stimuli were found to be α-ionone, l-octanol, skatol, isoamyl acetate, ethyl butyrate, coumarin, phenol and similar compounds.

Psychophysics of Taste Systems

The second type of study which has contributed to our understanding of the functional properties of oral chemoreceptor systems is human psychophysics, where verbal reports are taken on the taste properties of food and beverages and their chemical constituents. It is often possible for an individual to break a flavor complex down into a variety of distinguisable sensations. These sensations are end products of neural processing that are available to consciousness. Any natural food is of complex chemical composition and thus activates a wide variety of oral and nasal chemoreceptors. These flavor sensations may arise entirely from the oral cavity or require both oral and nasal stimulation.

Although it is common to assert that there are only four distinct taste sensations, even a casual introspection reveals that other oral sensations can be distinguished. As one may expect, flavor chemists have discovered that many separate oral sensations are required to reconstruct the flavors of foods and beverages. Some of these sensations have distinct oral loci from which they are elicited by specified types of chemical compounds, thus indicating that different neural systems are involved. Many of these sensations are difficult to typify verbally and also often have affective overtones. These sensations are the result of considerable peripheral and central neural processing and are only indirectly related to the peripheral neural pulse signals as discussed above. The type of sensation elicited and the locus of elicitation provide us with further measures of the functional properties of oral chemoreceptor systems.

Studies on human taste sensations confirm and extend our understanding of the types of chemical signals measured by these oral chemoreceptor systems. There are, for instance, several distinct sensations elicited by chemical stimulation of fungiform papillae innervated by the geniculate ganglion, indicating that a neural functional complexity similar to that described above for the cat, dog and goat underlies these human taste systems.

Table II summarizes some of the different types of taste sensations that can be elicited by chemical stimulation of the human oral cavity and the types of chemical compounds found to elicit them. In some cases it is possible to assign a sensation to a particular ganglion because of the locus of elicitation. The sensations in this table consist of only some of the more commonly elicited sensations and those with some degree of experimental specification. The amount of information available varies widely for the different sensations, and some of the sensations in this table may not be clearly differentiated from one another.

Four of the sensations commonly distinguished are the salty, sour. sweet, and bitter sensations. All of these sensations can be elicited from the fungiform papillae systems. The salty sensation is associated with relatively high concentrations of inorganic ions (16,17), particularly Na, K and Li. The sour sensation is elicited by various Brønsted acids with indications that proton donating nitrogen groups may be active at neutral pH (18-20). $Sweet_1$ and $bitter_1$ have been given subscripts to distinguish them from similar sensations elicitable from the back of the mouth. $Sweet_1$ is evoked by solutions of low concentrations of inorganic salts, sugars, and various nitrogen compounds, especially amino acids (21) such as L-hydroxyproline and L-alanine. $Bitter_1$ can be associated with hydrophobic amino acids (22) and alkaloids.

The sensation of pleasant is postulated on the basis of cat neurophysiology and human psychophysics. The pleasant sensation is assumed to arise from the stimulation of a small fiber geniculate ganglion system. The stimuli eliciting the pleasant sensation are lactones and other carbon-oxygen compounds (23). The general indistinctness of the pleasant sensation is assumed to be associated with the activation of extremely small fiber systems.

The sensations $sweet_2$ and $bitter_2$ can be distinguished because they are elicited from posterior oral loci by stimuli distinct from those acting on the front (17,24). Dihydrochalcones are active stimuli for $sweet_2$; and $bitter_2$ sensation is elicited by certain salts like $MgSO_4$, and probably various polyphenols. Additional "sweet" and "bitter" sensations could probably be distinguished. The sweet tasting proteins thaumatin and monellin have been found to maximally stimulate fungiform papillae on the lateral edge of the tongue as opposed to sucrose which stimulates the tip (25). Certain foods seem to elicit a bitter sensation localized to the foliate papillae.

"Umami" is the Japanese word used to describe the sensation elicited by compounds such as monosodium glutamate, sodium inosinate, sodium guanylate, and ibotenic acid (26-29). The umami sensation is sometimes translated as the sensation of "deliciousness". The possibility of more than one umami sensation exists, since the monophosphate nucleotides stimulate far back in the oral

Table II: Partial Summary of some Human Taste Sensations

Sensation	Locus	Stimuli	Receptor	Ganglion
1. Salty	Ant. Tongue, Palate	NaCl, KCl	Taste buds	Geniculate
2. Sour	Ant. Tongue, Palate	Malic Acid	Taste buds	Geniculate
3. $Sweet_1$	Ant. Tongue, Palate	L-Alanine, Fructose	Taste buds	Geniculate
4. $Bitter_1$	Ant. Tongue, Palate	L-Tryptophan	Taste buds	Geniculate
5. Pleasant	Ant. Tongue, Palate	Lactones	Taste buds	Geniculate
6. $Sweet_2$	Post. Tongue	Dihydrochalcone	Taste buds	Petrous
7. $Bitter_2$	Post. Tongue	$MgSO_4$, Phenolics	Taste buds	Petrous
8. Astringent	Oral Cavity	Theaflavin	Free Nerve	Trigeminal
9. Pungent	Oral Cavity	Capsaicin	Free Nerve	Trigeminal
10. $Umami_1$	Tongue	Monosodium Glutamate	?	?
11. $Umami_2$	Post. Mouth	IMP, GMP	?	?
12. Metallic	Tongue	Silver Nitrate	Taste buds (?)	Petrous

cavity. These umami sensations may be indistinct and difficult to characterize. It is probable that they arise from small fiber sensory systems. Compounds acting similarly to MSG are L-cysteine-S-sulfonic acid, L-homocysteic acid, and muscimol (22).

The metallic sensation arises from stimulation with certain metallic salts, such as silver nitrate, and is also associated with 1-octen-3-one (30). Metallic taste sensations may also arise with pathologies of the glossopharyngeal nerve.

Two of the sensations associable with trigeminal ganglion systems are the pungent sensation and the astrigent sensation. These sensations are elicitable from much of the oral cavity. The compounds eliciting the pungent sensation (31) are found in chillies, pepper, ginger, mustard and horseradish. Pungent compounds include piperine, capsaisin, gingerol, and sinigrin. Common stimuli for the astrigent sensation (32-34), include many polyphenols such as those found in fruits, cider, teas, wines, and beer.

Other chemical sensations associated with the trigeminal ganglion include temperature sensations such as the coolness (35) associated with menthol and the heat associated with capsaisin. Oral sensations elicited by chemical solutions also include tactile sensations such as smooth, dry, or powdery, and such disagreeable sensations as pain. Some of these sensations may represent different degrees of activation of a single system or the activation of several separate systems.

Many other oral chemical sensations may be distinguished, although in general there has been little study of them, their locus of elicitation or their chemical stimulus determinants. Sensations such as yeasty, nutty, soapy, fruity, papery, acrid, acid (as distinguished from sour) are often distinguished (36-38). Taste sensations with distinct hedonic tones (23,39) such as sweetish, creamy, coconut, peachy, and so forth are often elicited by phenolic compounds such as vanillin and various oxygen heterocycles (e.g. ethyl maltol), especially lactones. Disagreeable oral sensations such as burnt, stale, tainted and noxious are often reported. Many of these sensations doubtless include a nasal component.

Neurophysiological Correlates of Sensation

Different sensations may arise because of the activation of two distinct neural taste systems (e.g. cat group I and group II), or the differential activation of a single system. Differential activation of the same system could occur when different segments of an ordered population are activated by different chemicals or when one chemical compound excites the neural group and another inhibits.

On the basis of neurophysiology on the cat, dog and goat geniculate ganglia, the neurophysiological correlates of sour, $sweet_1$, and $bitter_1$ can be postulated with some degree of certainty. The sour sensation arises when a large fiber subgroup is activated. This large fiber system is a general mammalian system found in cats, dogs, humans, and goats. The stimuli activating this system are similar in all species studied. The system is highly responsive to acidic compounds, especially carboxylic and phosphoric acids. The stimuli for the activation of this system can be typified as Brønsted acids (14). At neutral pH, carboxylic and phosphoric acid groups are dissociated and hence nonstimuli. Most of the food eaten by animals is near neutrality and the most active compounds will be nitrogen Brønsted acids. Thus the cat and dog sour system will be stimulated by natural stimuli such as carnosine and histidine, and the goat by various plant compounds including alkaloids (Table I). The sourness of nitrogen compounds has been little studied, although meat (40) has been reported to elicit a sour sensation and L-histidine, an active stimulus in the dog and cat, elicits a small sour sensation (41) at a pH of 7.4.

The sweet sensation is associated with the activation of a geniculate ganglion neural group (cat group II, dog group A) and the bitter sensation with the inhibition of this same neural group (42). The equating of this carnivore system with the human sweet-bitter sensations is made on the basis of similarity of stimuli, especially amino acids (42). Amino acids that stimulate the cat and dog system tend to taste sweet, those that inhibit taste bitter. In both the cat and the dog this system is activated by NaCl and KCl compounds which taste sweet in low concentration. In the dog and the human, this system is activated by sugars. The properties of sweet and bitter nitrogen compounds have been described by Wieser et al. (43). It is evident that this system is specially modified for the different species, e.g., with distinctions among the types of amino acids stimulating the system. No amino acid sensitive system seems present in the goat, thus making the human more like the cat and dog than the goat.

As indicated elsewhere, there is evidence for a possible correlate of a pleasant sensation in a cat unit group, but this system has been little investigated in either the cat or in other species. Although the goat has a system that responds maximally to Na and Li salts, this system has not been seen in the carnivore. The chemical stimuli eliciting the human sensation of salty are salts in relatively high concentration, concentrations that in other species may stimulate more than one group.

Taste Compounds in Foods

In discussing natural taste compounds one faces a dilemma. On the one hand almost every compound occurring in nature is a possible taste compound, especially if it is at all water soluble. A vast number of possible taste compounds is thus arrayed before us. On the other hand, relatively few food compounds have been studied for their taste properties. Invariably, volatile compounds are assigned the role of olfactory stimuli, even though often the compounds must be put into the mouth to produce the flavor. In only a few cases has there been any recognition of taste properties of volatile flavor products. Therefore, at the present time there exists little knowledge of the taste activities of many natural flavor compounds; and in assembling any list of taste active compounds one often must operate partly on conjecture.

Man is capable of living on an all plant or all animal diet, although omnivory is most common in human societies. Animal foods may be divided into two major categories as far as we are concerned: vertebrate and invertebrate. Although invertebrate animals may play a significant part in the diet of many human societies, there has been little work on invertebrate taste chemistry (except for shellfish) from a human consumption standpoint. Shellfish taste is primarily due to inorganic ions, organic acids, amino acids and nucleotides (44).

Hunting and carnivory are found in humans, baboons and chimpanzees (45). For perhaps 2 million years, man has killed and eaten all varieties of vertebrates. As the ultimate carnivorous ape, man has exterminated most of the large animals of the earth and cut down most of the trees to cook them. Different animals, birds and eggs have widely different tastes; and, in addition, different parts of the body will have different tastes. Through studies on the flavor chemistry of raw fish and meat, much is known about vertebrate muscle flavor compounds (46-49). Prominent in meat and fish taste are inorganic salts, nucleotides, amino acids (especially sulfur amino acids), the dipeptides anserine and carnosine (which often occur in extremely high concentration), and various other compounds found in flesh such as taurine, thiamin, and organic acids. Egg flavor compounds are in large part similar to those found in meats. Milk flavor, however, largely derives from organic acids, simple phenolics, sugars and lactones.

Plant foods present another order of chemical complexity as compared to animal foods. The types of compounds present in plants are much more varied than those present in animal tissues (48, 50-56). The chemical composition of the seeds or fruit of a plant will be different from that of the roots, bark, leaves, or stems. The flavor of fruits is usually determined by compounds distinct from those functioning in the flavor of vegetables. Prominent in fruit flavor (39,51,57) for instance are sugars, alcohols, aldehydes, esters, organic acids and lactones (58). Vegetable flavors (48, 59-62), on the other hand, are usually attributed to various nitrogen and sulfur compounds, especially amino acids, nucleotides, and various nitrogen and sulfur heterocycles (63-66). Lactones (58), however, are prominent in celery and tomato flavor; and sugars are major factors in many root foods such as carrots and beets. Phenolic compounds (67,68) occur in all classes of vegetables and fruits. Mushroom flavor

(69) comes primarily from nitrogen and sulfur compounds, and some highly unusual compounds such as 1-octen-3-ol and muscimole may be active. Sulfur compounds are highly prominent in the flavor of garlic and onions (70,76), and are also important in the flavor of asparagus, tomatoes, and cabbage.

The taste of any food item would consist of the different oral chemical sensations elicited when this food is consumed. The types of sensations elicited would be a function of the classes of compounds present in the food (Table III), since different sensations will be evoked by different compounds. In Table IV are tabulated some of the taste sensations likely to be associated with different foods. Raw monkey meat would elicit a variety of sensations, including a salty, sour, strong $sweet_1$, $umami_1$, and $umami_2$. An apple from a natural nutritional ecosystem would likely elicit sensations of $sweet_1$, $bitter_2$, sour, $sweet_2$, astringent, and pleasant. Various other foods would elicit other sensations. The taste of any food would therefore be a composite of discrete taste sensations. Additional chemical sensations would derive from the olfactory and trigeminal systems. Some of these sensations may require both oral and nasal input, since oral and nasal chemoreceptor systems have been demonstrated to have converging input on brain stem neurons (72,73).

Alterations in Natural Nutritional Ecosystems

Man has made basic alterations in his nutritional ecosystem. Two of these changes have clearly been to intensify the flavor of the foods he eats. The first of these changes, fermentation, is seminatural in that the flavor compounds are quite likely to occur naturally in foods. Many of man's foods are subjected to fermentation before being consumed. Examples of such foods are alcoholic beverages such as wine and beer, many breads, pickles and condiments (e.g., kim chi and sour kraut), cheeses, many flavor sauces such as soy sauce and fish sauce (including anchovies), and beverages such as coffee, tea and cocoa. The flavor products developed during microbial fermentation depend in large part on the substrate and the microbe. Some of the common fermentation flavor products are alcohols, esters, fatty acids, various mono and dicarbonyls, phenolic compounds, and many lactones (74-77). In general, flavors from fermented foods are strong and complex.

The major man-induced change in the chemistry of his nutritional ecosystem is the production of flavor compounds by cooking foods; whereas microbial production of flavor compounds is seminatural, many heat produced compounds are not found in nature in any quantity. Although it has been speculated that food is cooked for hygienic purposes or to increase the nutritional value of foods eaten (certain starches are rendered edible by heating), most foods man presently eats can be eaten raw with little or no loss in nutritional value. With heating, in fact,

Table III: Simplified Summary of some of the Major Taste Active Compounds found in different foods.

Compound	Meat	Vegetables	Fruit	Roots	Seeds
Inorganic Ions	xx	x	x	x	x
Amino Acids	xxx	xx	x	xx	xx
Peptides and Proteins	xxx	xx		x	xx
Histidine Dipeptides	xxx				
Nucleotides	xxx	x		xx	xx
Amines	xx	x			
Sugars			xxx	xx	xx
Phenols - Simple		xx	xx		x
Hydroxy Compounds		x	xx		
Polyphenolic Compounds		xx	xx		x
Carbonyl Compounds		xx	xxx		x
Esters			xx		
Sulfur Compounds	xx	xx			x
Acids		xx	xxx		
Furans		xx	xx	x	x
Lactones		xx	xxx		
N, S Heterocycles	x	xxx	x		x

Table IV: Simplified Summary of the Taste of some Foods, uncooked.

TASTE SENSATION

Food	Salty	Sour	$Sweet_1$	$Bitter_1$	Pleasant	$Sweet_2$	$Bitter_2$	Astringent	Pungent	$Umami_1$	$Umami_2$	Metallic
Monkey	xx	x	xxx	x	x			x		xx	xxx	
Oyster	xx	x	xx	xx	x			x		xx	xxx	
Carrot	x		xxx		xx	x						
Apple		xx	xxx	x	xx	xx	xx	xxx				
Orange		xxx	xxx	x	xxx	xxx	xx	xx				
Tomato	x	xxx	xxx	x	xx	x	x	x		xx		x
Lettuce	x	x	x	xx		xx	xx	xx				
Wheat		x	xx	x	x					xx	xx	
Mushroom			xx	xx	xx					xx	xx	xx
Onion			xx		xx			xx				
Honey			xxxx	x	xx			x				
Ginger			xx	x	x		x	x	xxx			

the opposite may occur since many vitamins and amino acids are destroyed. Furthermore, many of the compounds produced by cooking have been reported to be detrimental to health (78,79). To any flavor chemist the reason for cooking foods is obvious: it gives them flavor. The variety of flavor compounds produced by heating foods is astronomical, especially when, as is common in preparing many dishes, foods of different chemical composition are heated together.

The flavor compounds produced by heating (63-66,79-82) include aldehydes, ketones, phenols, thiols, and a wide variety of sulfur nitrogen and oxygen heterocycles. Five and six member ring heterocycles are among the most flavor active compounds arising from cooking. Those with oxygen in the ring are most characteristic of plant foods, especially grain products, or of the lipid derived flavors of milk and fat. Popcorn and potato chips are flavored in part by these compounds. The complex flavors of coffee and chocolate are also in part derived from sulfur and nitrogen heterocycles, as are meat flavors. Heat produced heterocyclic sulfur and nitrogen compounds include pyridines, pyrazines, thiazoles, pyrroles, and thiophenes. For many of these compounds only olfactory sensations have been reported. They are, however, similar in structure to many heterocyclic compounds active neurophysiologically in the cat and dog and are therefore likely taste active in humans. Relatively little is known of the nutritional properties of many of these heat produced compounds. Many of them are of natural occurrence, although usually found at much lower levels. Others are unlikely to be encountered in a natural nutritional ecosystem, e.g. oxazoles and oxazolines (83).

In many cases fermentation and heating are involved together in food preparation. Thus both fermentation and heating are used in the production of tea, coffee, chocolate and alcoholic beverages. Although some natural foods can be typified by their simple mild tastes, these processed foods give rise to strong complex sensations. These complex sensations arise from the many taste active substances present. The taste of beer for instance, would result from compounds naturally present in grain and hops, such as amino acids, nucleotides, and various phenolic compounds and those produced by fermentation and heating such as alcohols, lactones and sulfur and nitrogen heterocycles (Figure 9).

Other major changes that man has instituted in the chemical composition of his natural nutritional ecosystem derive from agriculture and industry; and, unlike the changes incurred by fermentation and heating, many of these chemical changes have been detrimental for taste activity. In the agricultural revolution of the neolithic period, man substituted a nutritional ecosystem over which he had some control for one over which he had no control. Traditional agriculture approximated a natural nutritional ecosystem in that man was normally part of an intergrated system and food was produced by small diversified farms.

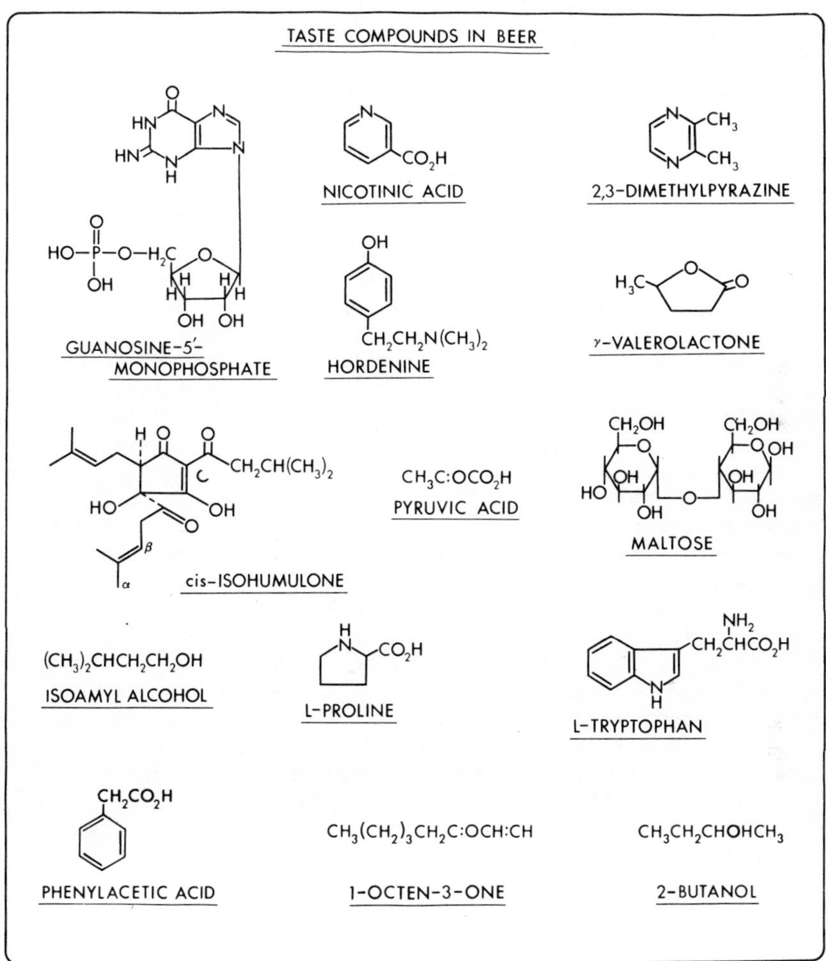

Figure 9. Some of the taste active compounds found in beer

Food was grown and selected in large part on the basis of taste. In the last 100 years however, changes have occurred in the methods of food production and distribution that are detrimental to good taste. Taste is no longer a factor in food production. Rather, agriculture is geared toward quantity production and various distribution needs. The chemical composition of the production system has been simplified by the elimination of most naturally occurring species and the farming of one or two crops. Fruits and vegetables are grown for yield and ease in transportation and are ripened by artificial means.

These changes have taken place largely without regard to taste (84). As a result many foods have lost taste, as exemplified by turkey and tomatoes (85). Several practices contribute to the general inedibility of American agricultural produce. Artificially ripening fruits results in a decrease in flavor compounds (Figure 10). In a natural nutritional ecosystem there is great chemical complexity as a result of the many species contributing to the flow of compounds. Many of these compounds such as amino acids and nucleotides are utilized by plants and hence cycled in the system. The reliance on a few fertilizers with high nitrogen content and a few simple compounds results in an imbalance in the ecosystem disrupting natural systems and changing chemical composition of food. The loss of flavor in onion and garlic will result with sulfur deficiency (86); a deficiency becoming more and more common in farms. In England there has occurred a shift from traditional growing of cider apples to intensive close packed orchards utilizing high nitrogen fertilizers. As a result flavor has declined markedly (87).

In fact it seems that agricultural chemistry as now practiced is inimicable to good taste. Besides the effect of industrial fertilizers upon food composition, the ultilization of chemical compounds for various agricultural purposes has been found to alter the chemical composition of our foods. The chemicals used to loosen oranges for mechanical picking, for instance, have been found to introduce novel chemicals with offtastes into the orange (88,89). Nematicides have also been reported to produce large changes in the chemical composition of tomatoes (90). The utilization of various chemical compounds for herbicides, fungicides, insecticides, and medication has introduced various new compounds into our foods (91,92). Many of these compounds are incorporated into the food we eat often in high concentration. Some of the toxicants present in foods (93) are endrin, DDT, toxaphine, aldrin and dieldrin, heptachlor, diazinon, parathion, chlorobenzilate, dithiocarbimate, dalapon, dimethoate and many other compounds employed for various purposes. Besides novel food compounds directly added by agriculture, many industrial compounds such as polychlorinated biphenols have found their way into our food supply. Some of the compounds of common occurrence in today's food are illustrated in Figure 11. These compounds and similar derived products are assumed to detract from the

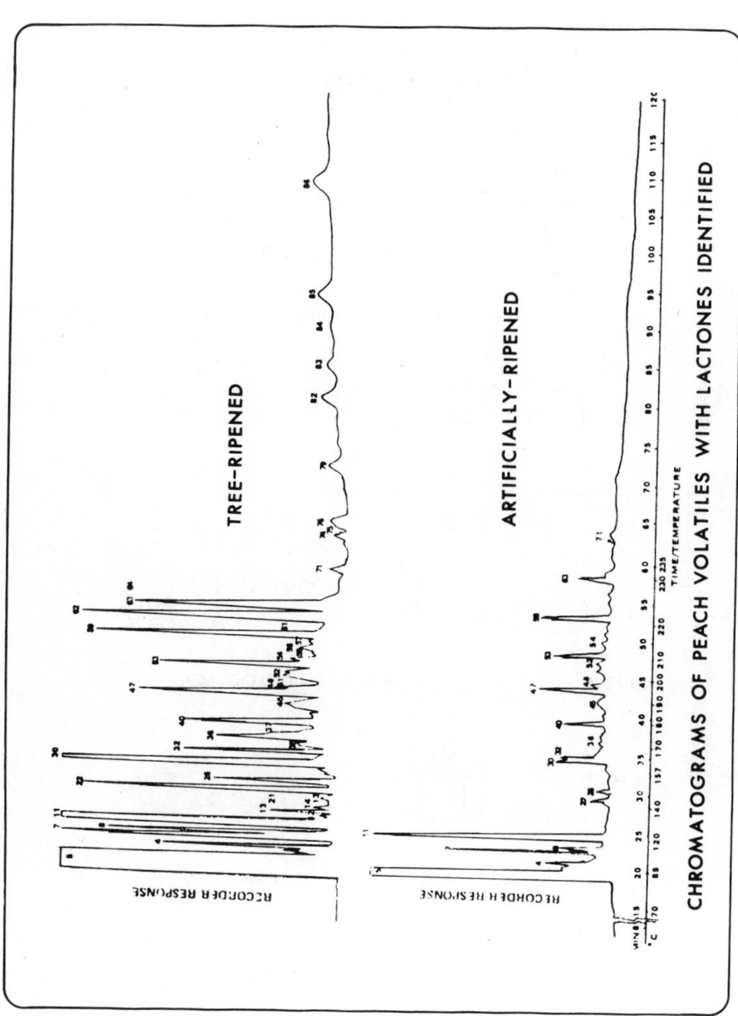

Journal of Food Science

Figure 10. The differences in the flavor volatiles of tree-ripened and artificially ripened fruit (95).

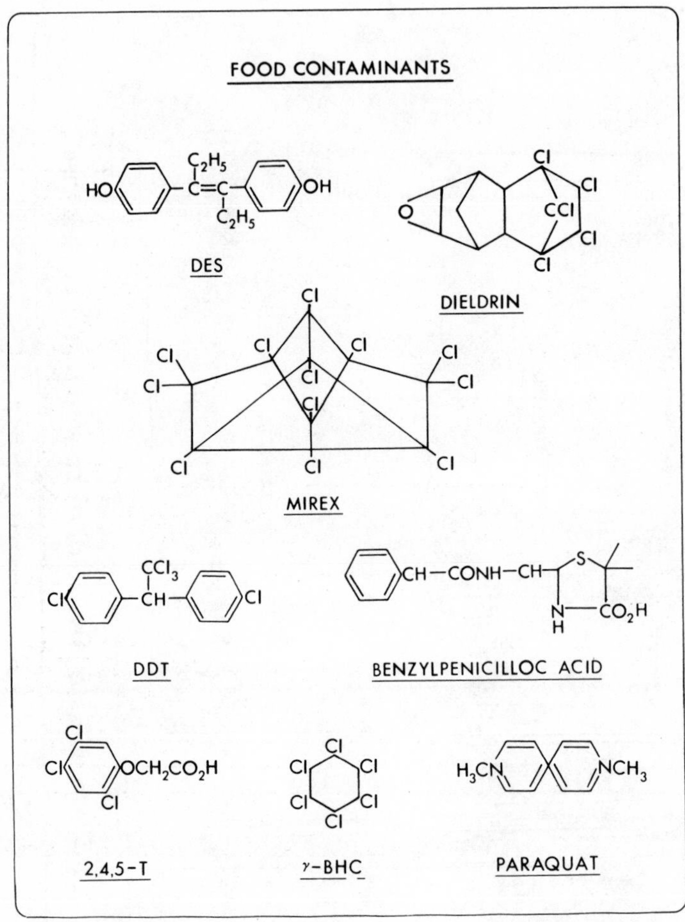

Figure 11. Some recent food additives

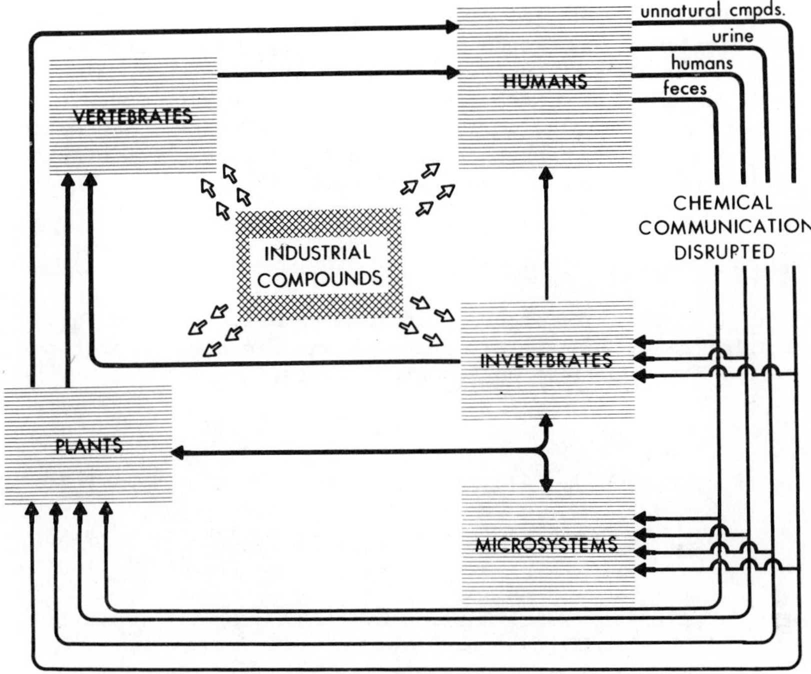

Figure 12. Schematic of present human nutritional ecosystem with diminished components

palatability of foods. Many of the foods selected over millenia for flavor have, in the last 20 to 30 years because of changes in their chemical composition, become bland or even objectionable in flavor. Some of the changes instituted in man's natural nutritional ecosystem are illustrated in Figure 12.

Concluding Statement

Taste in present day food production is often not a relevant variable, being secondary to such factors as yield, shipping, storage and appearance. Often the food industry gives the impression that food is something to which flavor can be added. The food future often projected is one in which we will be eating various processed foods such as flavored soy with infinite storage life. Flavor, however, is inherent in the natural chemical composition of the food and the only way to improve food flavor is by producing food similar in chemical composition to that found in the natural nutritional ecosystem within which the taste systems were designed to function.

Acknowledgements

We thank J. Lucke for computer programming and C. Holley for secretarial assistance. This research was financed in part by National Science Foundation Research Grants.

Literature Cited

1. Freeland, W.J., Janzen, D.H., Amer. Nat. (1974) 108: 269-289.
2. Swain, T. Ann. Rev. Plant Physiol (1977) 28: 479-501.
3. Cott, H.B., Proc. Zool. Soc., London (1946) 116: 371-524.
4. Cott, H.B., Proc. Zool. Soc., London (1953) 123: 123-141.
5. Ahern, E.M., In: "Medicine in Chinese Cultures" (A. Kleinman et al., Ed's), 91-113, U.S. Dept. Health Ed. Welfare, NIH, DHEW. Pub. No. 75-653, Wash. D.C., 1975.
6. Anderson, E.N., Anderson, M.L., In: "Medicine in Chinese Culture" (A. Kleinman et al., Ed's), 143-175, U.S. Dept. Health Ed. Welfare, NIH, DHEW. Pub. No. 75-653, Wash. D.C., 1975.
7. Boudreau, J.C. and Tsuchitani, C., "Sensory Neurophysiology", Van Nostrand Reinhold Co., N.Y., 1973.
8. Andres, K.H., Arch. Oto. -Rhino.-Laryng. (1975) 210: 1-41.
9. Graziadei, P.P.C., In: "Olfaction and Taste III" (C. Pfaffmann Ed.), 315-330, Rockefeller University, N.Y., 1969.

10. Murray, R.G., In: "Handbook of Sensory Physiology". IV. part 2, 31-50.
11. Shimamura, A., Tokunaga, J., Toh, H., Arch. Hist. Jap. (1972) 34: 52-60.
12. Takeda, M., Hoshino, T., Arch. Hist. Jap. (1975) 37: 395-413.
13. Miller, I.J., J. Comp. Neurol. (1974) 158: 155-166.
14. Boudreau, J.C., White, T., In: "Flavor Chemistry of Animal Foods" (R.W. Bullard, Ed) 102-128, Amer. Chem. Soc., Wash. D.C., 1978.
15. Kashiwagura, T., Kamo, N., Kurihara, K., Kobatake, Y. Comp. Biochem. Physiol. (1977) 56: 105-108.
16. Moncrief, R.W. "The Chemical Senses", L.Hill, London (1967).
17. Skramlik, E.v., Physiologie des Geschmaksinnes, In: "Handbuch der Physiologie der Niederen Sinne", Georg Thieme Verlag, Leipzig, 1926.
18. Beets, M.G.J., "Structure Activity Relationships in Human Chemoreception", Applied Science Pub., Ltd., Barkin, 1978.
19. Boudreau, J.C., Nelson, T.E., Chem. Sen. Flav. (1977) 2: 353-374.
20. Amerine, M.A., Pangborn, R.M., and Roessler, E.B., Principles of Sensory Evaluation of Food, Acad. Press, N.Y., 1965.
21. Kirimura, J., Shimizu, A., Kimizuka, A., Ninomiya, T., and Katsuya, N., J. Agri. Food Chem. (1969) 17: 689-695.
22. Ney, K.H. In: "Natürliche und Synthetische Zusatzstoffe in der Nahrung des Menschen. (R. Ammon and J. Holló, Eds) 131-143.
23. Arctander, S., "Perfume and Flavor Chemicals" S. Arstander, Publisher, Elizabeth, N.J. (1969).
24. Hall, M.J., Bartoshuk, L. M., Cain, W.S., Stevens, J.C. Nature (1975) 353: 442-443.
25. Van der Well, H., Arvidson, K. Chem. Sen. Flav. (1978) 3: 291-297.
26. Kuninaka, A., In: "Chemistry and Physiology of Flavors" (H.W. Schultz, E.A. Day and L.M.Libbey, Eds.), Avi. Publishing Co., Westport, Conn. pp 515-535 (1967).
27. Terasaki, M., Fujita, E., Wada, S., Takemoto, T., Nakajima, T., Yokobe, T., Jrnl. Jpn. Soc. Food Nutr. (1965) 18: 172-175.
28. Terasaki, M., Fujita, E., Wada, S., Takemoto, T., Nakajima, T., Yokobe, T., Jrnl. Jpn. Soc. Food Nutr. (1965) 18: 222-225.
29. Yamaguchi, S. In: "Olfaction and Taste VI" (J. Le Magnen and P. Mac Leod, Eds) p 493, Information Retrieval, Wash. D.C., 1978.
30. Meilgaard, M.C., MBAA Tech. Quart. (1975) 12: 151-168.

31. Govindarajan, V.S., CRC Crit. Rev. Food Sci. Nutr. (1977) 9: 115-225.
32. Lea, A.G.H., Arnold, G.M., J. Sci. Fd. Agric. (1978) 29: 478-483.
33. Herrmann, K., Deutsche Lebensmit.-Rundschau (1972) 68: 105-141.
34. Charalambous, G., Katz, I. (Eds) "Phenolic, Sulfur and Nitrogen Compounds in Food Flavors" Am. Chem. Soc., Wash.D.C., 1976.
35. Watson, H.R., In: "Flavor: Its Chemical, Behavioral and Commercial Aspects" (C.M. Apt, Ed.), 31-50, Westview Press, Col., 1978.
36. Clark, R.G., Nursten, H.E., Int. Flav. Fd. Add. (1977) 8: 197-201.
37. Clapperton, J.F., Dalgliesh, C.E. and Meigaard, M.C., J. Inst. Brew. 82: 7(1976).
38. Holn, E., Solms, J., Lebensmwiss. u.-Technol. (1976) 8: 206-211.
39. Furia, T.E. and Bellanca, N., "Fenaroli's Handbook of Flavor Ingredients", 2nd Ed., CRC Press, Cheveland, Ohio, (1975).
40. Caul, J.F., In: "Chemistry of Natural Food Flavors", 152-167, Quartermaster Food and Container Institute for the Armed Force, Wash. D.C., 1957.
41. Ninomiya, T., Ikeda, S., Yamaguchi, S., Yoshikara, T., Rept. 7 th Sensory Evaluation Symposium, JUSE, pp 109-123, 1966.
42. Boudreau, J.C., In: "Flavor of Foods and Beverages Chemistry and Technology" (G. Charalambous and G.E. Inglett, Eds) 231-246, Academic Press, N.Y., 1978.
43. Wieser, H., Jugel,H., Belitz, H.D., Z. Lebensm. Unters. Forsch. (1977) 164: 277-282.
44. Hashimoto, Y., In: "The Technology of Fish Utilization" (R.Kreuzer, Ed), 57-61, Fish News (Books), London, 1965.
45. Hamilton, W.J., Busse, C.D. Bioscience (1978) 28: 761-766.
46. Konosu, S., Watanabe, K., Shimizu, T., Bull. Jap. Soc. Sci. Fish. (1974) 40: 909-915.
47. Mabrouk, A.F., In: "Phenolic, Sulfur, and Nitrogen Compounds in Food Flavors" (G. Charalambous and I. Katz, Eds) 146-183, Am. Chem. Soc., Wash. D.C., 1976.
48. Solms, J., In: "Gustation and Olfaction an International Symposium" (G. Ohloff and A.F. Thomas Eds) 92-110, Academic Press, N.Y., 1971.
49. Solms, J., In: "Aroma-und Geschmacksstoffe in Lebensmitteln" (J. Solms and H. Neukom, Eds) Forster Verlag AG, Zurich, 1967, 199-221.
50. Herrmann, K., J. Fd. Technol. (1976) 11: 433-448.
51. Herrmann, K., Qual. Plant. (1976) 25: 231-246.
52. Herrmann, K., Z. Lebensm. Unters.-Forsch. (1974) 155:

220-233.
53. Lee, C.Y., Shallenberger, R.S., Vittum, M.T. Food Sciences (Geneva, N.Y.) No. 1, 1970.
54. Linner, K., Qual. Plant. (1973) 23: 251-262.
55. Schmidtlein, H., Herrmann, K.Z. Lebensm. Unters.-Forsch. (1975) 159: 139-148.
56. Smith, T.A., Phytochem. (1975) 14: 865-890.
57. Nursten, H.E. In: "Sensory Properties of Foods" (G.G. Birch, J.C. Brennan, and K.J. Parker, Eds), 151-166, Appl. Sci. Pub., Barkin, 1977.
58. Maga, J.A., Crit. Rev. Fd. Sci. Nutr., (1976) 8: 1-56.
59. Salunkhe, D.K., Do, J.Y., Crit. Rev. Fd. Sci. Nutr., (1976) 8: 161-190.
60. Schutte, L., Crit. Rev. Fd. Tech. (1974) 4: 457-505.
61. Virtanen, A.I., Phytochem. (1965) 4: 207-228.
62. Maga, J.A., CRC Crit. Rev. Fd. Sci. Nutr., (1978) 10: 373-403.
63. Maga, J.A., CRC Crit. Rev. Fd. Sci. Nutr., (1976) 7: 147-192.
64. Maga, J.A., CRC Crit. Rev. Fd. Sci. Nutr., (1975) 6: 241-270.
65. Maga, J.A., CRC Crit. Rev. Fd. Sci. Nutr., (1975) 6: 153-176.
66. Maga, J.A. and Sizer, C.E., J. Agr. Food Chem., 21 (1973) 22-30.
67. Murray, K.E., Whitfield, F.B., J. Sci. Fd. Agric. (1975) 26: 973-986.
68. Maga, J.A., CRC Crit. Rev. Fd. Sci. Nutr., (1978) 10: 323-372.
69. Dijkstra, F.Y., Wiken, T.O., Z. Lebensm. Unters.-Forsch. (1976) 160: 255-262.
70. Freeman, G.G., Whenham, R.J., J. Sci. Fd. Agric. (1975) 26: 1869-1886.
71. Whitaker, J.R., Adv. Fd. Res. (1976) 22: 73-133.
72. Van Buskirk, R.L.v., Erickson, R.P. Neurosc. Let. (1977) 5: 321-326.
73. Van Buskirk, R.L.v., Erickson, R.P., In: "Olfaction and Taste VI", p 206, Information Retrieval, Wash.D.C., 1977.
74. Haymon, L.W., In: "Lipids as a Source of Flavor" (M.K. Supran, Ed.), 94-115, Am. Chem. Soc., Wash.D.C., 1978.
75. Litman, I., Numrych, S., In: "Lipids as a Source of Flavor" (M.K. Supran, Ed.), 1-17, Am. Chem. Soc., Wash. D.C., 1978.
76. Tressl, R., Apetz, M., Arrieta, R., Grunewald, K.G., In: "Flavor of Foods and Beverages Chemistry and Technology" (G. Charalambos and G.E. Inglett, Eds.) 145-168.
77. Artman, N.R., Adv. Lip. Res. (1969) 7: 245-330.
78. Commoner, B., Vithayathil, A.J., Dolara, P., Nair, S., Madyastha, P., Cuca, G.c., Science (1978) 201: 913-916.
79. Chang, S.S., Peterson, R.J., Ho, C.T., In: "Lipids as a

Source of Flavor" (M.K. Supran, Ed.) 18-41, Am. Chem. Soc., Wash.D.C., 1978.
80. Schutte, L., In: "Phenolic, Sulfur, and Nitrogen Compounds in Food Flavors" (G. Charalambous and I. Katz, Eds) 96-113, Am. Chem. Soc., Wash.D.C., 1976.
81. Mussinan, C.J., Wilson, R.A., Katz, I., Hruza, A., Vock, M.H., In: "Phenolic, Sulfur and Nitrogen Compounds in Food Flavors", (G. Charalambous and I. Katz, Eds), 133-145, Am. Chem. Soc., Wash.D.C., 1976.
82. Wilson, R.A., Agr. Fd. Chem. (1975) 23: 1032-1037.
83. Maga, J.A., J. Agr. Fd. Chem. (1978) 26: 1049-1050.
84. Dirinck, P., Schreyen, L., Schamp, N., Agr. Fd. Chem. (1977) 25: 759-763.
85. Whiteside, T., "The New Yorker", 1977, Jan 14, 36-61.
86. Freeman, G.G., Whenham, R.J., Int. Flav. (1976) 7
87. Lea, A.G.H., Beech, F.W., J. Sci. Fd. Agic. (1978) 29: 493-496.
88. Moshonas, M.G., Shaw, P.E., J. Agric. Fd. Chem. (1977) 25: 1151-1153.
89. Moshonas, M.G., Shaw, P.E., J. Agric. Fd. Chem. (1978) 26: 1288-1290.
90. Bajaj, K.L. Mahajan, R., Qual. Plant. (1977) 27: 335-338.
91. Oehme, F.W., Toxicology (1973) 1: 205-215.
92. Menn, J.J., Still, G.G., CRC Crit. Rev. Toxicol. (1977) 5: 1-21.
93. Salunkhe, D.K., Wu, M.J., CRC Crit. Rev. Fd. Sci. Nutr., (1977) 9: 265-324.
94. Miller, I.J., In: "Food Intake and Chemical Senses" (Y. Katsuki et al, Eds) 173-185, University Park Press, Baltimore, 1978.
95. Do, J.Y., Salunkhe, D.K., Olson, L.E., J. Food Sci. (1969) 34: 618- 621.

RECEIVED August 9, 1979.

2

The Umami Taste

SHIZUKO YAMAGUCHI

Central Research Laboratories, Ajinomoto Co., Inc., Kawasaki, Japan

The characteristic taste of monosodium glutamate and 5'-ribonucleotides is called "umami" in Japanese. It plays a predominant role in the flavor of foods, such as meats, poultry, fish and other sea foods, dairy products, or vegetables. The taste was first discovered by Ikeda (1908)(1), and has been studied by a large number of researchers from different points of view (refer to, e.g. 2-9).

We have systematically investigated the umami taste using psychometric procedures. In this paper, a part of our studies will be outlined.

Umami Substances

Most typical umami substances are divided into two series of compounds. One is a group of L-α-amino acids represented by monosodium glutamate (MSG) (Table I) (10-15), and another is that including 5'-ribonucleotides and their derivatives, represented by disodium 5'-inosinate (IMP) or disodium 5'-guanylate (GMP)(Table II) (16, 17, 18, 19). The latter group of substances have only very

Table I. Umami Substances Related to MSG

	Relative umami intensity
Monosodium L-glutamate H_2O	1
Monosodium DL-*threo*-β-hydroxy glutamate H_2O	0.86
Monosodium DL-homocystate H_2O	0.77
Monosodium L-aspartate H_2O	0.077
Monosodium L-α-amino adipate H_2O	0.098
L-Tricholomic acid (*erythro* form)[a]	5-30
L-Ibotenic acid[a]	5-30

Table from Yamaguchi *et al.* (38).
[a]From Terasaki *et al.* (14, 15).

0-8412-0526-4/79/47-115-033$05.00/0

Table II. Umami Substances Related to IMP

Substance (Disodium salt)	Relative potency of umami
5'-Inosinate.7.5H_2O	1
5'-Guanylate.7H_2O	2.3
5'-Xanthylate.3H_2O	0.61
5'-Adenylate	0.18
Deoxy 5'-guanylate.3H_2O	0.62
2-Methyl-5'-inosinate.6H_2O	2.3
2-Ethyl-5'-inosinate.1.5H_2O	2.3
2-Phenyl-5'-inosinate.3H_2O [a]	3.6
2-Methylthio-5'-inosinate.6H_2O	8.0
2-Ethylthio-5'-inosinate.2H_2O	7.5
2-Ethoxyethylthio-5'-inosinate [a]	13
2-Ethoxycarbonylethylthio-5'-inosinate [a]	12
2-Furfurylthio-5'-inosinate.H_2O [a]	17
2-Tetrahydrofurfurylthio-5'-inosinate.H_2O [a]	8
2-Isopentenylthio-5'-inosinate (Ca) [a]	11
2-(β-Methallyl)thio-5'-inosinate [a]	10
2-(γ-Methallyl)thio-5'-inosinate [a]	11
2-Methoxy-5'-inosinate.H_2O	4.2 (3.7)[a]
2-Ethoxy-5'-inosinate [a]	4.9
2-*i*-Propoxy-5'-inosinate [a]	4.5
2-*n*-Propoxy-5'-inosinate [a]	2
2-Allyloxy-5'-inosinate (Ca) 0.5H_2O [a]	6.5
2-Chloro-5'-inosinate.1.5H_2O	3.1
N^2-Methyl-5'-guanylate.5.5H_2O	2.3
N^2, N^2-Dimethyl-5'-guanylate.2.5H_2O	2.4
N^1-Methyl-5'-inosinate.H_2O	0.74
N^1-Methyl-5'-guanylate.H_2O	1.3
N^1-Methyl-2-methylthio-5'-inosinate	8.4
6-Chloropurine riboside 5'-phosphate.H_2O	2
6-Mercaptopurine riboside 5'-phosphate.6H_2O	3.4
2-Methyl-6-mercaptopurine riboside 5'-phosphate. H_2O	8
2-Methylthio-6-mercaptopurine riboside 5'-phosphate.2.5H_2O	7.9
2',3'-*o*-Isopropylidene 5'-inosinate	0.21
2',3'-*o*-Isopropylidene 5'-guanylate	0.35

From Yamaguchi *et al.* (38).
[a]From Imami *et al.* (19).

weak tastes. It is notable, however, that they synergistically increase the umami of the former substances (17, 20). In addition to these two groups, some peptides have been reported to have tastes similar to MSG (21, 22). Some researchers regard the taste of succinic acid (23, 24) or theanine (25) as umami, although their taste qualities are considerably different from that of MSG.

Fundamental Taste Properties of Umami

Threshold Value. The threshold values of umami and the other four taste substances have been reported by many researchers (e.g. 1, 26, 27, 28, 29). However, because of the measurement conditions are different, precise comparisons with one another are difficult. We have measured the detection thresholds of MSG and the four taste substances simultaneously as carefully as possible, using a single panel under identical experimental conditions (9, 30). The panel was composed of 30 laboratory members between the ages of 20 and 40. Triangle tests were used, where each triangle consisted of two samples of pure water and one sample of a test solution. The panelists were asked to select the odd sample. A series of triads were presented in descending order. The lowest concentration which could be significantly distinguished from pure water was obtained for each test substance (Table III).

Table III. Detection Thresholds for Five Taste Subsatnces (n = 30)

MSG	Sucrose	Sodium chloride	Tartaric acid	Quinine sulfate
0.012	0.086	0.0037	0.00094	0.000049

Concentrations given as g/100ml. From Yamaguchi and Kimizuka (9).

The detection threshold for MSG was as low as 0.012 g/100ml or 6.25×10^{-4}M. It was higher than that of quinine sulfate or tartaric acid, lower than that of sucrose and almost the same as that of sodium chloride in the molar concentration. Some umami substances have lower thresholds than that of MSG.

Subjective Intensity Scale for Umami. Thresholds do not always express the relative potency of different taste stimuli, because the intensity of taste does not increase with concentration in the same manner for each substance.

Several kinds of taste intensity scales have been established for the four tastes. Typical examples are the gust scale by Beebe-Center (31) and the τ scale by Indow (32). In order to deal with the umami on the same basis with the four tastes, we newly established a new subjective taste intensity scale for umami as well as for the four tastes (9, 30). Six solutions for each of the five taste substances were prepared. The taste intensity of each solution was rated by the 30 panel members. The panelists kept 10 ml of the sample in their mouths for 10 seconds. Then they were asked to assess the intensity of the taste on a 100 point scale with 0 being no discernible taste and 100 the highest intensity. The panelists rated all the samples twice. The results are shown in Figure 1 using mean values of a total of 60 ratings.

The relationship between the concentration and the perceived taste intensity of MSG was logarithmically linear like those of the four common tastes, although the slope for MSG was somewhat less steep than the others. It means that Weber-Fechner's law holds for all of the five taste substances. The relation of the taste intensity (S) to the concentration (x) can be expressed by

$$S = \alpha \log_2 (x/\beta),$$

where α represents the increase of taste intensity by doubled concentration and, β, the concentration at the point of intersection of the extrapolated line with the concentration axis, seen in Figure 1. This equation was applied to the five taste substances and the results were:

MSG $S_M = 9.69 \log_2 (x_M /0.0195)$,

Sucrose $S_S = 14.98 \log_2 (x_S /0.873)$,

Sodium chloride $S_{SC} = 15.50 \log_2 (x_{SC}/0.0943)$,

Tartaric acid $S_T = 14.45 \log_2 (x_T /0.00296)$,

Quinine sulfate $S_Q = 14.16 \log_2 (x_Q /0.000169)$,

where x is given in terms of g/100 ml.

In this experiment, only the intensity of taste was rated and the quality of taste was disregarded. Consequently, the same value of S in the above mentioned equations represents the same intesnity of taste. Beebe-Center defined the unit of taste intensity as gust. One gust means the taste intensity of 1% sucrose solution. However, the gust scale is not always convenient because it does not define the upper limit of the scale.

In our scale, the unit of S was adjusted so that the value of S is zero at the concentration β and 100 at the saturated sucrose concentration at 20°C (89.27g/100ml).

Interactions between Umami and the Four Tastes. Interaction of tastes is another important problem in the study of phenomena of tastes. In order to examine the effect of umami on the four common tastes, the influence of MSG on the thresholds of the four tastes has been examined by several researchers (27, 28, 33), but the results are conflicting. In order to clarify the issue, the thresholds of the four taste substances were measured again in 5mM solution of MSG or IMP (9, 30). The panel and experimental conditions were exactly the same as that in the aforementioned experiment. The detection threshold of quinine sulfate was slightly raised by the presence of 5mM of IMP. The threshold of tartaric acid was considerably raised by both umami substances, no doubt because of the change in pH. No effect was observed on the thresholds of sucrose and sodium chloride (Table III and IV).

Table IV. Detection Thresholds for the Four Taste Substances in Solution of Umami Substance

(n = 30)

Base solution	Sucrose	Sodium chloride	Tartaric acid	Quinine sulfate
0.094g/100 ml MSG[a]	0.086	0.0037	0.0019	0.000049
0.26g/100 ml IMP[a]	0.086	0.0037	0.03	0.0004

Concentrations given as g/100ml.
[a] 5mM.
From Yamaguchi and Kimizuka (9).

The Synergistic Effects of Umami Substances

Quantitative analysis of the synergistic effect. When fructose and sucrose are mixed together, the sweetness of the mixture becomes slightly greater than the sum of the sweetness of the separate substances (34). Such phenomenon is called the synergistic effect. A clear and precise definition of the synergistic effect along with several numerically treated examples has been presented elsewhere (34). The magnitude of the synergistic effect between the two groups of umami substances is unparalleled. Figure 2 shows the relationship between the intensity of umami and the proportion of IMP in the mixture of MSG and IMP (35). The total concentration was kept constant at 0.05 g/100ml and the proportion of IMP was varied from 0 to 100%. Since the umami intensities of the samples on both extremes are very weak and almost the same, the curve would have proven to be horizontally linear if the synergistic effect had been absent. The symmetric curve illustrates the remarkable synergistic effect. In this curve, the intensity of umami at its maximum is equivalent to that of 0.78g/100 ml of MSG alone. The mixture is 16 times as strong as that of MSG. This amplification factor is concentration-dependent, and becomes higher with increasing concentration.

The synergistic effect between MSG and IMP can be expressed by means of the following simple equation:

$$y = u + 1200\ uv \qquad (1)$$

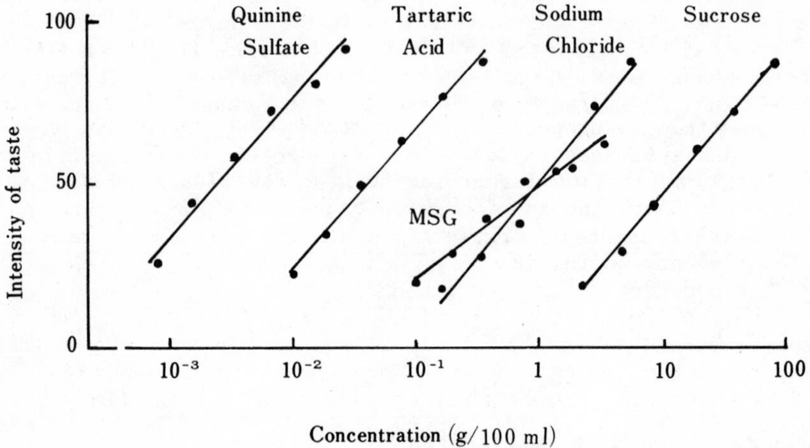

Figure 1. Relationship between taste intensity and concentration (9)

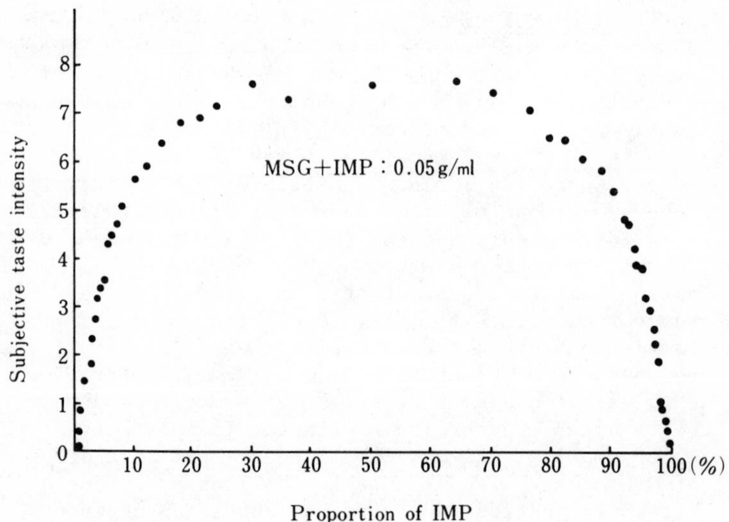

Journal of Food Science

Figure 2. Relationship between umami intensity and mixing ratio of MSG and IMP (35)

where u and v are the respective concentrations of MSG and IMP in the mixture and y is the equi-umami concentration of MSG alone(35).

The synergistic effect can be demonstrated between any combination of substances in Table I and Table II; and the intensity of umami can also be expressed by an equation essentially equal to equation (1) (36, 37, 38). The intensities of all substances in Table I are always proportional to that of MSG. Therefore, u'g/100ml of any substance in Table I is replaceable with $\alpha u'$g/100ml of MSG. The constant α for each substance is listed in Table I. On the other hand, the tasting activities of all nucleotides in Table II are consistently proportional to that of IMP. Hence, v'g/100ml of any nucleotide is replaceable with $\beta v'$g/100ml of IMP. The constant β for each nucleotide is listed in Table II. Therefore, the umami intensity of the mixture of any combination of substances in Table I and Table II can be calculated by substituting $\alpha u'$ for u and $\beta v'$ for v. Since the interrelationships within each series of substances are additive, the intensity of umami of the mixture of two or more different L-α-amino acids and two or more nucleotides can be calculated by substituting the product sums, $\Sigma\alpha_i u_i$ and $\Sigma\beta_j v_j$ for u and v, respectively, in equation (1).

Mathematical Consideration of the Synergistic Effect. Generalizing the above-mentioned results, we can introduce a concept of an umami taste space. The umami solution of any combination of the two groups of umami substances can be expressed as a point in a space of 2-dimensions, say, U, defined as follows: Let $U = \{(u, v)\}$ be a set of ordered pairs of non-negative real numbers. Given two elements $Y_1 = (u_1, v_1)$ and $Y_2 = (u_2, v_2)$, we define their sum by

$$Y_1 + Y_2 = (u_1 + u_2,\ v_1 + v_2),$$

and the scalar multiple by

$$\lambda Y = (\lambda u,\ \lambda v), \qquad \lambda : \text{non-negative real number.}$$

The absolute value of Y is defined by

$$|Y| = u + \gamma uv \qquad \gamma : \text{positive constant.}$$

To be concrete, if we take MSG and IMP as the standards for both groups of substances, u and v represent the totals of acidic amino acids and nucleotides in terms of the concentrations of MSG and IMP, respectively. In our taste space, the sum means combining the components of two solutions, and the scalar multiple means concentrating or diluting of a solution. The absolute value means the intensity of umami of a solution in terms of the concentration of, say, MSG alone.

Then the following inequality holds:

$$
\begin{aligned}
|Y_1 + Y_2| &= (u_1 + u_2) + \gamma(u_1 + u_2)(v_1 + v_2) \\
&= (u_1 + \gamma u_1 v_1) + (u_2 + \gamma u_2 v_2) + \gamma(u_2 v_1 + u_1 v_2) \qquad (2) \\
&\geqq |Y_1| + |Y_2|
\end{aligned}
$$

It is of interest to compare this space with the Euclidean Space R_n whose points and vectors are ordered n-tuples of real numbers,

$$X = (x_1, x_2, \cdots, x_n).$$

The absolute value of the vector $X = (x_1, x_2, \cdots, x_n)$ is defined to be the number which is called the "length" of the vector,

$$\|X\| = (x_1^2 + x_2^2 + \cdots + x_n^2)^{\frac{1}{2}}.$$

As well known, for any two vectors $X_1, X_2 \in R_n$, the inequality

$$\|X_1 + X_2\| \leqq \|X_1\| + \|X_2\| \qquad \text{(Trianglar law)} \qquad (3)$$

holds. However, the direction of the ordering-symbol in inequality (2) is just opposite to that of inequality (3). Inequality (2)means that in our umami space in which the synergistic effects occur, intensity of umami is always strengthened, by combining the components of two or more solutions, to more than the sum of the intensities of the original solutions.

Factors Affecting the Synergistic Effect

In order to determine the possible enhancement or suppression of the synergistic effects of umami substances, it was examined whether equation (1) held or not in the presence of various other taste substances.

Effects of the Four Taste Substances. The concentration of MSG equivalent in the umami intensity to an MSG-IMP mixture (point of subjective equality) was determined both in pure water and in the four taste solutions. The results are shown in Table V. The equi-umami concentration of MSG obtained in the presence of each of the four taste substances was almost the same as that in pure water. Thus the synergistic effects of umami substances were seen to be unaffected by the four taste substances.

Effects of Amino Acids. According to equation (1), the intensity of umami of 0.075 g/100ml MSG-IMP mixture containing 4% IMP is equivalent to that of 0.33 g/100ml MSG in pure water. The intensities of umami of these two samples were compared in various amino acid solutions using a paired sample test. In each pair, the panelists were asked to indicate which umami is stronger. No significant difference was recognized except for the basic amino acids, histidine and arginine (Table VI). Since these basic amino acids do not enhance the umami of MSG, they are assumed to suppress the synergistic effect of IMP. The suppression was dependent on both the pH value and buffer capacity of the solution involved. When the pH value of histidine solution was adjusted to 5 to 6.5, the synergistic effect was recovered. The synergistic effect was not affected by a low concentration of histidine, even at a high pH value (Table VII). Neutral amino acids and other buffer substances showed similar effects depending on the pH value and buffer capacity of the solution involved.

Table V. Equi-umami Concentration of MSG to MSG-IMP Mixture in the Four Taste Solutions

Base Solution		MSG-IMP mixture		Equi-umami
Compound	Concn. (g/100ml)	Prop. IMP (%)	Concn. (g/100ml)	concn. MSG (g/100ml)
Sucrose	5.0	4	0.075	0.31
		12	0.05	0.33
Sodium chloride	1.0	4	0.075	0.36
		12	0.05	0.36
Tartaric acid	0.05	4	0.075	0.36
Quinine sulfate	0.002- 3	4	0.075	0.34
(Pure water)		4	0.075	0.33
		12	0.05	0.36

Effects of Umami Substances on Flavors of Foods

Flavor Profiles of Foods Added MSG. The effects of MSG on a variety of foods were qualitatively and quantitatively investigated using the Semantic Differential (9, 39). The aim of the study was to make clear how people in general, not specialists in food science, respond to the flavor changes of foods.

In these experimental procedures, descriptive terms were collected first. Eight kinds of foods with and without additional MSG or containing different concentrations of broth or stock were served to 180 persons. The subjects expressed freely their impressions of the flavors of the samples using their own terminology. Out of approximately 500 expressions obtained, 32 pairs of the terms frequently expressed were selected except the terms

Table VI. Comparison of Umami Intensity between MSG-IMP Mixture and MSG in Amino Acid Solutions

(n = 25 or 50)

Base solution		No. judgement	
Compound	Concn. (g/100ml)	Mixture[a] > MSG[b]	MSG > Mixture
Alanine	0.95	9	16
Arginine	0.25	11	39**
Arginine·HCl	0.25	10	15
Glycine	1.65	13	12
Histidine	0.80	7	43***
Histidine·HCl	0.038	11	14
Isoleucine	0.35	15	10
Leucine	0.50	12	13
Lysine·HCl	0.64	11	14
Methionine	0.52	10	15
(Pure water)		11	14

[a]0.075g/100ml MSG-IMP mixture containing 4% IMP.
[b]0.33g/100ml MSG.
, * : significant at 1%, and 0.1% level, respectively.

Table VII. Equi-umami Concentration of MSG to MSG-IMP Mixture[a] in Histidine Solutions

(n = 50)

Concn. histidine (g/100ml)	pH[b]	Equi-umami concn. MSG (g/100ml)
0.8	9.5	0.09
0.8	8.5	0.16
0.8	7.5	0.21
0.8	6.5	0.38
0.8	5.0	0.33
0.2	9.5	0.16
0.05	9.5	0.31
(Pure water)	-	0.33

[a]0.075g/100ml MSG-IMP mixture containing 4% IMP.
[b]Adjusted with aqueous HCl.

of umami or MSG taste since the purpose was to clarify the flavor profile of MSG itself. The evaluation sheet was prepared based on these terms (Figure 3).

In the main experiment, 16 dishes were evaluated using a panel consisted of 300 ordinary people with the panel size for each session being 25 to 50. Each panelist was given a test sample added MSG and a control. The panelists were asked to compare the test sample against the control, and to evaluate the flavor of the test sample checking off the point on each scale of the evaluation sheet.

The data obtained were analyzed by multivariate analysis. The paired terms were classified into five major groups according to flavor functions, and some highly correlated terms were united. A typical profile chart is shown in Figure 4.

MSG, when added to beef consommé, had no effect on the aroma. It increased the overall taste intensity, but its effects on the intensities of saltiness, sweetness, sourness, and bitterness were very small. Although the term "umami" was intentionally not used in the profile test, it may be easily supposed that the quality of the taste increased here is umami. The addition of MSG increased the characteristics of the flavor, i.e., continuity, mouth fullness, impact, mildness and thickness of the flavor of the beef consommé. It also increased the meatiness of the flavor. Thus MSG increased the overall preference of the beef consommé.

The same profile of MSG was observed for many other foods, such as soup, meat, poultry, fish, egg and vegetable dishes.

Doubling the concentration of beef consommé gave the same pattern of change in the flavor profile of beef consommé as did the addition of MSG, but additionally increased the intensities of aromas and the four tastes (Figure 4).

From the extensive profile tests mentioned above, the effects of MSG on foods were summarized as follows:

(1) MSG has no effect on the aroma of food.
(2) MSG increases the total taste intensity of food. The quality of the taste brought about by MSG is different from the four tastes.
(3) MSG enhances certain flavor characteristics of food: continuity, mouth fullness, impact, mildness, and thickness.
(4) MSG enhances the specific flavor of food, e.g. meatiness of soups.
(5) MSG has a flavor effect similar to broth or stock, although MSG has no effect on aroma.
(6) MSG increases the preference or palatability of food.

The effects of sodium chloride and sucrose were also examined for reference. The beef consommé mentioned above contained 0.8g/100ml NaCl. Increasing the NaCl level to 1.2 g/100ml changed only the saltiness of the food and decreased its palatability (Figure 5). However, an increase of NaCl from 0.2 g/100ml to

FLAVOR PROFILE OF [FOOD NAME]

Date ____________
Name ____________

DIRECTIONS: Mark each line in the place that best expresses your feelings of SAMPLE B compared with SAMPLE A.

			certainly -2	slightly -1	almost same 0	slightly 1	certainly 2	
1.	Whole aroma	/ weak						strong
2.	Meaty aroma	/ weak						strong
3.	Aroma derived from(...)	/ weak						strong
4.	Whole aroma	/ bad						good
5.	Meaty flavor	/ weak						strong
6.	Flavor derived from(...)	/ weak						strong
7.	Flavor of spice	/ weak						strong
8.	Whole taste	/ weak						strong
9.	Salty taste	/ weak						strong
10.	Salty taste	/ rough						smooth
11.	Sweet taste	/ weak						strong
12.	Sour taste	/ weak						strong
13.	Bitter taste	/ weak						strong
14.	Meaty taste	/ weak						strong
15.	Taste drived from(...)	/ weak						strong
16.	Oily or fatty	/ weak						strong
17.	Foreign flavor	/ weak						strong
18.	Contimuity	/ short						long
19.	Simple							Complex
20.	Watery							Concentrated
21.	Mouthfullness	/ weak						strong
22.	Development	/ narrow						broad
23.	Flat							Body
24.	Light							Heavy
25.	Poor							Rich
26.	Thin							Thick
27.	Harsh							Mild
28.	Crude							Aged
29.	Balance	/ bad						good
30.	Punch	/ weak						strong
31.	Unfavorable							Tasty
32.	Palatability	/ bad						good

Raven Press

Figure 3. Evaluation sheet for flavor profile test (9)

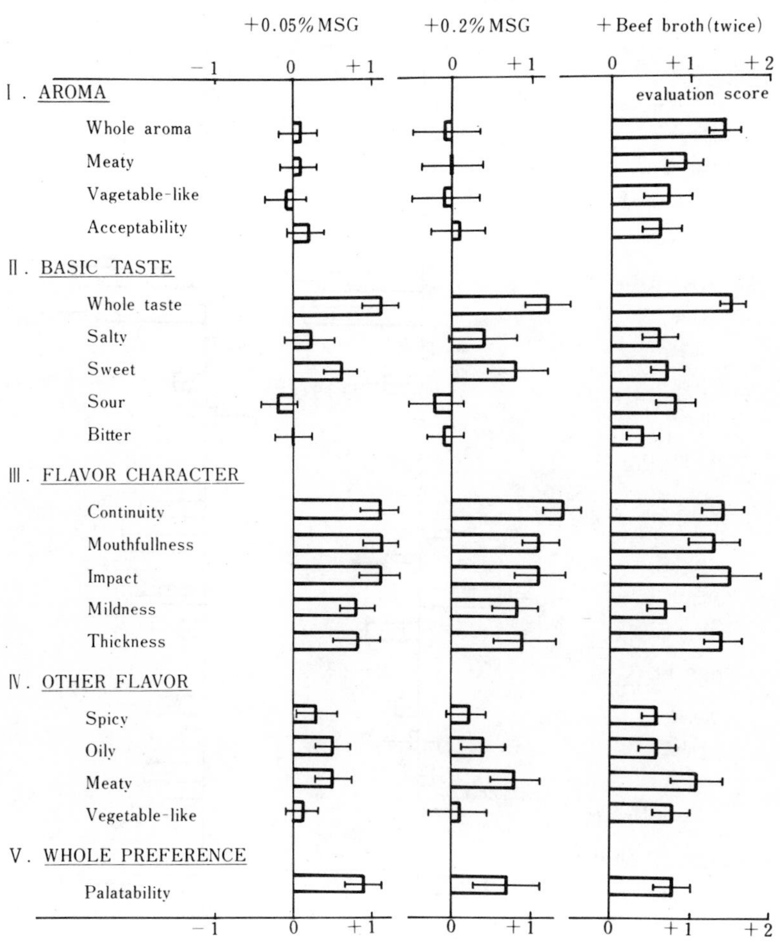

Figure 4. Effects of MSG and beef broth on flavor of beef consommé (9)

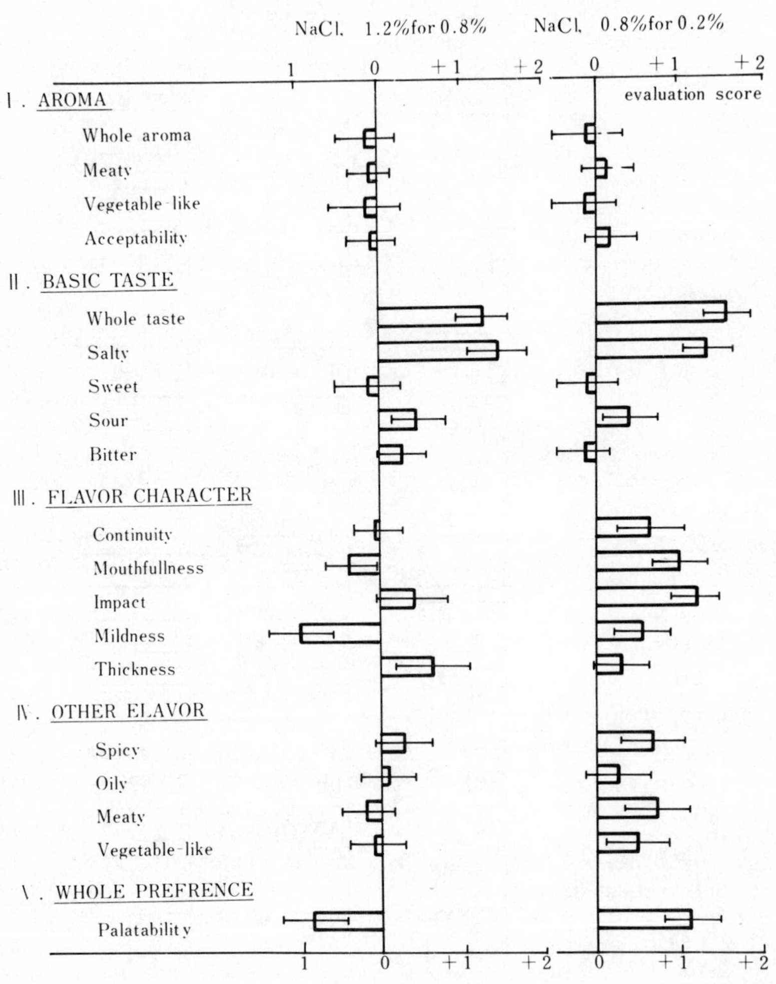

Raven Press

Figure 5. Effect of NaCl on flavor of beef consommé (9)

0.8 g/100ml enhanced palatability and increased the flavor characteristics of continuity, mouth fullness, impact, and so on (Figure 5).

In comparing the different sucrose contents of bavarian cream between 5% and 10%, the latter elicited larger evaluation scores of continuity, mouth fullness, impact, mildness and thickness, as well as increased sweetness (Figure 6).

Thus, in some cases, both NaCl and sugar not only increase their intrinsic tastes, but also enhance the flavor charactor measures.

Preference of Food and Content of Umami Substnces. In order to show the relationship between the preference of food and its content of umami substances, the value of y in equation (1) was calculated substituting u and v by the values of chemical analysis for glutamate and nucleotides, respectively, of each food presented in the flavor profile tests. A part of the results is shown in Table VIII.

The y value of beef consommé with no additional MSG was 0.15. By increasing beef broth concentration, the y value increased to 0.59 with a preference score of 0.80. The addition of 0.05g/100ml MSG to the beef consommé increased the y value to 0.91 by the synergistic effect of both IMP and GMP, which were contained naturally in the food itself, and gave 0.85 of the preference score. The addition of a small amount of MSG gave a large y value to this food, as did the increase of beef broth concentration, and increased the preference score. The close relationship was observed between the preference of food and the content of umami substances, in terms of the y value, whether they are added intentionally to food or contained naturally in food.

Relationship between Palatability and Umami. Yamanaka *et al.* (40) collected words expressing "palatability". They did this by asking people to write down their definition of palatability, excluding appearance, aroma and texture. From the total of 1900 expressions obtained, 38 of them were selected as important. The similality between each pair of the expressions was measured on a 5-point scale using a mass panel. The data obtained were analyzed by principal component analysis and cluster analysis. As a result, concrete expressions of palatability were classified into the following five groups:

(1) Full of body, concentrated, broad development, mild, aged, etc.
(2) Sharp, hot, spicy, pungent, etc.
(3) Refreshing, cool, clear, etc.
(4) Oily, fatty, greasy, etc.
(5) Refined, high grade, modern, etc.

Apart from the Group 5, our study demonstrated that the umami substances contribute mainly to the Group 1.

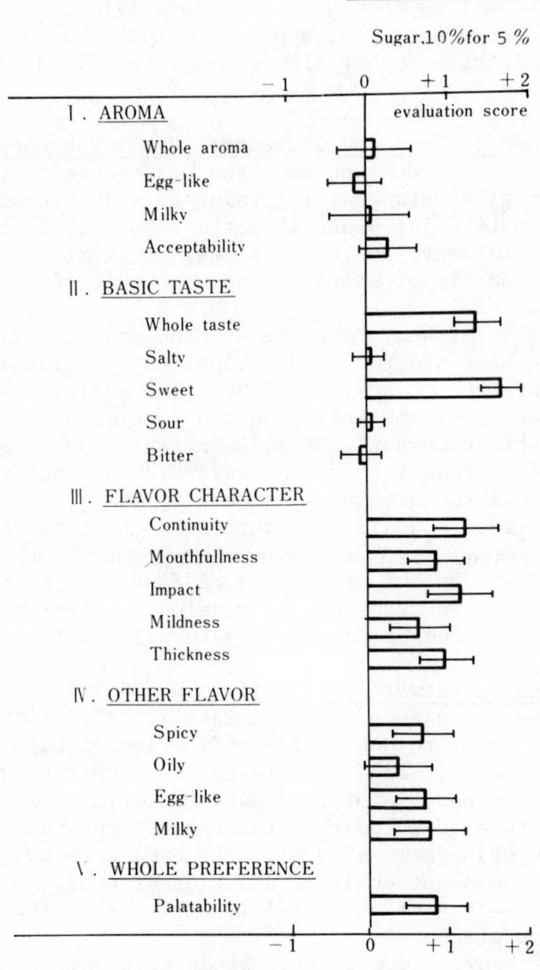

Figure 6. Effect of sugar on flavor of bavarian cream (9)

Table VIII. Relationship between y Value and Palatability of Food

Item	Control (A)				Test sample (B)			Difference of B to A
	Chemical analysis			y	Additional		y	
	MSG (%)	IMP (%)	GMP (%)			(g/100ml or %)		Preference score
Beef consommé	0.010	0.0113	0.0002	0.15	Beef broth	-	0.59	0.80
	"	"	"	"	MSG	0.05	0.91	0.85
	"	"	"	"	"	0.10	1.67	0.62
	"	"	"	"	"	0.20	3.19	0.67
	"	"	"	"	"	0.40	6.08	0.03
Chicken consommé	0.023	0.0097	0.0005	0.31	MSG	0.05	1.01	0.54
Cream of chicken soup	0.010	0.0023	0.0006	0.05	"	0.17	0.83	0.85
Chicken noodle soup	0.008	0.0014	0.0001	0.02	"	0.18	0.63	0.87
Cream of vegetable soup	0.026	0.0005	0.0002	0.06	"	0.05	0.16	0.49
	"	"	"	"	Chicken broth	-	0.21	0.71
Vichyssios	0.011	n.d.	0.0003	0.02	MSG	0.18	0.30	0.58
Onion soup	0.012	"	n.d.	0.01	"	0.50	0.51	0.85
Cream of tomato soup	0.122	"	0.0006	0.32	"	0.30	1.11	0.17
	"	"	"	"	Chicken broth	-	0.92	0.08
Japanese miso soup	0.074	n.d.	n.d.	0.07	MSG	0.30	0.37	0.56

From Yamaguchi and Kimizuka (9).

In this work, we have clarified psychometrically both the fundamental taste properties of umami in itself and its flavor effects on foods.

Umami is a kind of taste quality different from the traditional four tastes. Umami substances added to food increase not only their own taste, "umami", but also the flavor characteristics such as continuity, mouth fullness, impact, mildness and thickness. Thus they increase the palatability of foods.

Acknowledgement

The auther wishes to express her appreciation to Drs. J. Kirimura, Y. Komata and A. Kimizuka for their continuing guidance and encouragement. The author is also grateful to Drs. J.C. Boudreau and Y. Sugita for their critical reading of the manuscript.

Literature Cited

1. Ikeda, K. J. Tokyo Chem. Soc., 1909, 30, 820.
2. "Flavor and Acceptability of Monosodium Glutamate" (Proceedings of the First Symposium on Monosodium Glutamate); The Quartermaster Food and Container Institute for the Armed Forces and Associates, Food and Container Institute, Inc.: Chicago, 1948.
3. "Monosodium Glutamate: A Second Symposium" (Proceedings of the Second Symposium on Monosodium Glutamate); The Research and Development Associates, Food and Container Institute, Inc.: Chicago, 1955.
4. Kuninaka, A.; Kibi, M.; Sakaguchi, K. Food Technol., 1964, 18, 287.
5. Amerine, M.A.; Pangborn, R.M.; Roessler, E.B., Ed. "Principals of Sensory Evaluation of Food"; Academic Press: New York, 1965; p. 115.
6. Motozaki, S., Ed. "Chemical Seasonings"; Korin-shoin: Tokyo, 1969.
7. Ogata, K.; Kinoshita, S.; Tsunoda, T.; Aida, K., Ed. "Microbial Production of Nucleic Acid-Related Substances"; Wiley: New York, 1976; p. 299.
8. Kare, M.R.; Maller, O., Ed. "The Chemical Senses and Nutrition"; Academic Press: New York, 1977; p. 343.
9. Filer, L.J. Jr.; Garattini, S.; Kare, M.R.; Reynolds, W.A., Ed. "Glutamic Acid: Advances in Biochemistry and Physiology"; Raven Press: New York, 1979.
10. Akabori, S. J. Jpn. Biochem. Soc., 1939, 14, 185.
11. Keneko, T.; Yoshida, R.; Katsura, H. J. Chem. Soc. Jpn., 1959, 80, 316.
12. Kaneko, T. J. Chem. Soc. Jpn. ,1938, 59, 433.
13. Kaneko, T.; Yoshida, R.; Takano, I. The Abstract Papers of the

14th Annual Meeting of the Chemical Society of Japan (Tokyo), 1961; p. 305.
14. Terasaki, M.; Fujita, E.; Wada, S.; Nakajima, T.; Yokobe, T. J. Jpn. Soc. Food Nutr., 1965, 18, 172.
15. Terasaki, M.; Wada, S.; Takemoto, T.; Nakajima, T.; Fujita, E.; Yokobe, T. J. Jpn. Soc. Food Nutr., 1965, 18, 222.
16. Kodama, S. J. Kokyo Chem. Soc., 1913, 34, 751.
17. Kuninaka, A. J. Agric. Chem. Soc. Jpn.,1960, 34, 489.
18. Yamazaki, A.; Kumashiro, I.; Takenishi, T. Chem. Pharm. Bull., 1968, 16, 338.
19. Imai, K.; Marumoto, R.; Kobayashi, K.; Yoshioka, Y.; Toda, J.; Honjo, M. Chem. Pharm. Bull., 1971, 19, 576.
20. Toi, B.; Maeda, S.; Ikeda, S.; Furukawa, H. The Abstract Papers of the General Meeting of the Agricultural Chemical Society of Japan (Tokyo), 1960.
21. Kaneko, T. Kagaku to Kogyo (Chemistry and Chemical Industry), 1971, 24, 846.
22. Arai, S.; Yamashita, M.; Fujimaki, M. Agric. Biol. Chem., 1972, 36,1253.
23. Takahashi, T. J. Brew. Soc. Jpn., 1912, 7(12), 7.
24. Aoki, K. J.Agric. Chem. Soc. Jpn., 1932, 8.867.
25. Sakato, Y. J.Agric. Chem. Soc. Jpn., 1949, 23, 262.
26. Knowles, D.; Johnson, P.E. Food Res., 1941, 6, 207.
27. Lockhart, E.E.; Gainer, J.M. Food Res., 1950, 15,459.
28. Mosel, J.N.; Kantrowitz, G. Am. J. Psychol., 1952, 65, 573.
29. Pfaffmann, C. The sense of taste. In "Handbook of Physiology" (Magoun, Ed.); Am. Physiol. Soc; Washington, D.C., 1959; p. 507.
30. Le Magnen, J.; MacLeod, P., E. "Olfaction and Taste VI"; Information Retrieval: London, Washington D.C., 1977; P. 493.
31. Beebe-Center, J.G. J. Psychol., 1949, 28, 411.
32. Indow, T. Jpn. Psychol. Res., 1966, 8, 136.
33. Pilgrim, F.j.; Schutz, H.G.; Peryam, D.R. Food Res., 1955, 20, 310.
34. Yamaguchi, S.; Yoshikawa, T.; Ikeda, S.; Ninomiya, T. Agric. Biol. Chem., 1970, 34, 187.
35. Yamaguchi, S. J. Food Sci., 1967, 32, 473.
36. Yamaguchi, S.; Yoshikawa, T.; Ikeda, S.; Ninomiya, T. J. Agric. Chem. Soc. Jpn., 1968, 42, 378.
37. Yamaguchi, S.; Yoshikawa, T.; Ikeda, S.; Ninomiya, T. Agric. Biol. Chem., 1968, 32, 797.
38. Yamaguchi, S.; Yoshikawa, T.; Ikeda, S.; Ninomiya, T. J. Food Sci., 1971, 36, 846.
39. Osgood, C.E.; Suci, G.J.; Tannenbaum, P.E. "The Measurement of Meaning"; University of Illinois Press: Chicago, 1957.
40. Yamanaka, M.; Okayasu, S.; Tanaka, H. Unpublished data.

RECEIVED August 7, 1979.

3

Pungency: The Stimuli and Their Evaluation

V. S. GOVINDARAJAN

Central Food Technological Research Institute, Mysore, India 570013

The quality that distinguishes an appetising meal, which one wants to eat, from a collection of cooked components that satisfy merely one's nutritional requirements, is flavor. Aroma, and the traditionally accepted four tastes, sweet, sour, salt, and bitter, are part of most foods and they are further developed or altered during cooking. Even so, most cooked foods are still considered insipid, and require food additives, such as spices, herbs and potentiators, to boost the flavour. The important contributions of spices and herbs to sensory qualities, other than aroma, are the new dimensions, pungency and astrigency. Apart from imparting a specific altered aroma, they increase the pre- and post-ingestional cues which are important in increasing awareness and appreciation of the food, and thereby lead to increased ingestion (1). Thus, the notion of food flavour - which is commonly understood as aroma and taste - should be expanded to cover the contribution of pungency.

Pungency, an appraisal of the term

A definition of 'pungency' is required, if we are to understand the perceived impression, and attempt to estimate the same. The dictionaries define the term as 'a stinging, irritating, or caustic property' (Oxford) and 'keenness, sharpness, poignancy' (Webster). These are either too general, or they are rather unhappy choices of descriptors which connote some undesirable characteristics. The word pungent is also used to denote acrid odors, and, in the general literature, it is used almost as a synonym of 'hot' and 'irritant'. However, with respect to food, for which the term 'pungency' should really be used, it assumes a desirable quality. When used at an optimal usage level, the spices have - apart from aroma - a 'mouth-watering' quality. Together with aroma, pepper, ginger and chillies provide the 'piquant' stimulus response that leads to a greater acceptance and higher intake of food. This desirable nature of the reaction, I believe, identifies 'pungency' as different from the other descriptors used to define the attribute, such as irritant, stinging, caustic, etc., all of

0-8412-0526-4/79/47-115-053$10.00/0

which have undesirable connotations.

We could also look at the attribute from the stimulus-sensation angle. Moncrieff (2), in his treatise on the Chemical Senses, has classified pungency under the common chemical senses, grouping it along with stimuli eliciting burning (hot), irritation, lachrymatory or cough-provoking sensations. I would like to submit that all these properties are different sensations, though some chemicals might elicit more than one of these sensations.

Many examples can be found which show these differences. Ammonia, while imparting a highly irritant and caustic sensation cannot be called pungent. The lachrymatory substance of onion, the propanol S-oxide, is not pungent in the mouth, but produces an itching reaction in the nose. Sulphurdioxide, a cough-provoking chemical, produces no pungent feeling in the mouth, but only a catch at the throat. While a drinking concentration of alcohol could bring about the sensation of warmth in the throat, and even a burning sensation at higher concentrations, it can never be called pungent. There are other compounds in some spices and herbs which produce a tingling sensation on the tongue, but this again is distinct from the typical sensation of pungency elicited by the characteristic components of spices, such as capsaicin, piperine and gingerol. Singleton and Noble recently (3) described pungency as a hot, penetrating, burning sensation in the mouth which at lower levels may be warm, spicy, sharp or harsh. This implies that all these varying sensations, are merely concentration effects.

Such a view may be accepted only in so far as the terms warm, burning, penetrating and irritating are understood in their generally accepted meanings; but ordinary phenol, which elicits these responses, will at no concentration be called pungent. Also, typical pungent compounds at all levels, from threshold to higher levels, are clearly identified as pungent. These specific spice components are perceived as pungent only when taken in the mouth, while contact with other parts, such as the skin, is felt as burning or irritation; in the nose as watering, and in the eyes as lachrymatory. Whatever be the ultimate reason for the different responses to the same compound at different centres of perception, the distinct response as pungency when taken in the mouth is associated with specific compounds, both natural and synthetic, and restricted to some structural characteristics (see later). These properties/characteristics should be considered significant enough to afford pungency a status of its own among the different sensory attributes of food.

The next point for consideration is whether pungency is a purely gustatory modality, or whether it is also an olfactory modality. There is a vast and growing literature on the interrelation of the senses, though the extent of their mutual influences are not well understood. It is, however, a common experience that once taken into the mouth, foods and beverages elicit at least dual, usually triple, sensory effects due to the interconnection of the throat and nasal cavity, and the innervation of the receptors disposed in relation to one another. The senses of taste and cutaneous sensibilities in the mouth, and of olfaction in the nose appear to act simultaneously in response to a stimulus. However, these could be independently experimented upon using various devices. As pungent compounds can be felt in the mouth at low threshold levels, and not by smell at those levels, pungency may be classified along with the traditional taste modalities. Additionally, as with other tastes, there is little or no variation of quality, but only variation in intensity, and an optimal concentration for acceptance. The accepted definition of the term taste, 'one of the senses, the receptors for which are located in the mouth and are activated by a large variety of different compounds' (4), will accomodate the quality of pungency as well. Though most investigators now limit gustatory qualities to four, namely, sweet, sour, saline, and bitter, Ayurveda in India, and also the sciences of other cultures, recognised six kinds of taste; the above-mentioned four as well as pungent and astringent.

I would like to make a passing reference to the distinctness of the chemoreceptors. The receptor for the four basic tastes is now recognised to be a neural complex of specialized, elaborately organised cells, distributed on the tongue in a localized fashion. The receptors for other chemicals are reported to be the so-called free nerve endings, and are found throughout the oral cavity. Receptors located on other parts of the body responding to touch, temperature, pressure, are all reported to have distinctive nerve endings, while for irritation and pain no recognisable receptor bodies have been identified (5). Would further work with the more advanced electron microscopes yield further information and show up differences between the free nerve ends in the mouth and those outside the oral cavity? Recent work in neurophysiology and electron microscopy has discovered a cold receptor for former 'free nerve endings' (5a). Would the free nerve endings of recognised taste receptors respond to the application of specific compounds eliciting the pungency response? The extensive work with capsaicin by Jansco-Gabor and colleagues (5b) has shown pronounced desensitization against all kinds of chemical stimuli; it is most likely that receptors belonging to the slowest conducting C fibres are involved in its effect. Alternatively, is the difference in the response to the specific pungency stimuli, within the mouth and outside, related to the localisation of the neural information at the higher centres of transmission?

The widespread use of the terms 'pungent or bite component', 'capsaicin pungency' or 'sensory pungency' has tended to be confusing. To recognise the cause and effect relationship clearly, the stimulant should be identified only by its common or chemical name, and the sensory response by the term pungency.

Estimation of Pungency - Problems

When pungency is considered as an important gustatory attribute of food, an estimation of this perceived pungency is necessary for validating any quality control procedure of the raw material, or of the prepared food. Even when the chemical compounds that are responsible for eliciting this sensation are known in each raw material and instrumental methods are available for their estimation, the subjective estimation of pungency is necessary to establish a correlation as a guide to the use-level. Sensory evaluation of food for any perceived food quality had in the past been considered an art and in general very variable; but over the last two or three decades, its contribution to the success of processed foods has been realised. As in the development of any analytical method, the causes of variation and sources of error have been studied and much objectivity has been introduced into the practical methods of sensory evaluation (6,7). The common use of the term 'objective' for physico-chemical measurements and 'subjective' for evaluation by humans should also be discouraged, since there are elements of 'objectivity' and 'subjectivity' in both measuring systems. It is preferable to specifically state the instrumental measure and the sensory modality used. Many food laboratories are practising sensory evaluation by panels, in the place of one or a few experts, in order to arrive at decisions on the raw material or finished product quality, process change, storage life etc., (8). It is heartening to see that many aspects of sensory evaluation of foods are regularly discussed at meetings; and that institutions concerned with standards and specifications are, through their special committees, actively (9) discussing and codifying definitions, details of methods, giving guidance in the selection of methods of evaluation and analysis needed to answer specific questions, and evolving a uniform method of reporting sensory evaluation results.

It is important to consider, at this juncture, if there is any quality dimension in pungency. In the earlier literature, there are some references to the quality of pungency by descriptions such as instant, vanishing, persistent, etc. These observations on natural and synthetic compounds were, however, made by different people, and by procedures which were not standardised. An extreme example of the importance of correct and detailed observation is the misinterpretation of the lack of pungency in piperine crystals isolated from pepper extracts, while the mother liquor had high pungency; that mistake led to years of search by many workers to identify the non-existent isomers of piperine in pepper (see later). Todd et al have recently observed (10) that capsaicin, dihydrocapsaicin, and nordihydrocapsaicin gave rapid pungent sensations located in the back of the palate and throat, while homocapsaicin and homodihydrocapsaicin tended to produce prolonged pungent sensations of low intensity located in the mid-mouth and mid-palate regions. They have also found that these effects are probably related to the solubility of these components in polar solvents, higher for the first three components and lower for the latter two homologs with longer alkyl side-chains.

It appears to be clear that, in common with other gustatory attributes, there are no quality differences in pungency, and that the perceived differences are explained by intrinsic intensity, concentration, and solubility of the different stimuli responsible for pungency. As mentioned earlier, in any food situation, the interaction of other sensory modalities could result in seemingly apparent variations in pungency quality of different stimuli, but such differences are not real. We have carefully looked into this aspect by comparison of isolated, pure, total pungent components (free from aroma) of the spices, chillies, pepper, and ginger. At concentrations slightly above their individual thresholds, the different, pure stimuli are indistinguishable from one another even by trained panelists. At higher levels, some distinctions could be made; e.g. quick and strong for capsaicinoids, 'low' for gingerol, 'slow' for shogaol, and 'slow and lasting' for piperine. These differences could, however, be related to their known intrinsic, intensity differences and solubilities in water. It would be preferable to measure their solubility in saliva, and also to find out whether there are differences in their rates of penetration to the receptors.

It was early recognised that pungency evaluation needs to be done as a dilution method. Scoville (11), as early as 1912, determined the pungency of capsicum by determining the greatest dilution at which the definite perception of 'bite' (pungency) would be recognised (the recognition threshold), and expressed it as the reciprocal of the dilution. This figure has since been called the Scoville heat units. Though apparently simple, several factors interfere in this subjective judgement. For example, variation in acuity of panelists; aroma lowering the thresholds and thereby biasing the judgement to higher values than the real; carry-over of pungency and adaptation effect biasing judgement of subsequent samples to lower values; other psychological anticipatory errors now recognised in sensory testing; non-standardisation of procedures and preparation of samples for dilution testing; and lack of pure reference compounds had all contributed to poor reproducibility and resultant scepticism about the validity of the test. With the standardisation of threshold tests generally for other taste and aroma stimuli, pungency testing has also improved, and we can now get reliable and reproducible measurements of pungency.

Standard methods for evaluation of pungency:

The American Spice Trade Association (ASTA) (12), adopted, in 1968, an official method (21.0) for pungency evaluation for capsicum, which appears to be an adaptation of procedures then in use by some flavour houses. These methods were probably developed as a routine quality control procedure, on samples obtained usually from particular sources, to check whether the usual level of pungency was being maintained; they depended on serial dilution around the expected value, and accepting the level at which three out of five, trained, in-house panelists found pungency. The ASTA procedure improved on this by mentioning that the test should be performed as a threshold test and that carry-over effects should be avoided by palate clearing and allowing time between samples; a dilution table to cover a wide range of Scoville values from 100 to 15 x 10^5, expected for chilli and its oleoresins was also provided. However, the method gave no recommendation on concentration differences in the dilution series, and retained the single figure value agreed upon by three out of five judges. The processing and treatment of data are not statistically acceptable, nor useful for correlative work with chemical methods of estimations of the stimuli. These are probably the reasons for the poor reproducibility of the method recently reported (10). This procedure has, however, been adopted by the British and by the International Standards Organisation.

All these aspects have been recently studied in two publications (10,13), which standardized the dilution test for pungency and clearly established correlations with the estimates of total, and even individual, capsaicinoids. These papers also review the earlier attempts at standardisation. Two approaches are possible: use of a fairly homogenous panel to determine the threshold pungency response due to the stimuli; or use of a general panel to determine the average threshold of the panel for the stimuli. The second value will have wider applicability in use situations; but the first value should be useful for correlative work. The two methods approach one another when panels are screened and trained to avoid all bias factors, and when carefully planned dilution levels, details of panel procedure, treatment of data, and expression of results are adopted. In fact, the results published in the two independent studies (10,13), show close values, 17±0.9 million for natural capsaicinoids and 16.1±0.6 million for pure capsaicin and dihydrocapsaicin.

The method standardized in our laboratory can be summarised by the following steps, each of which significantly improves reproducibility and precision.

i) Preparation of sample and preliminary testing, when judges are familiarised with the recognition of pungency bias factors, the test procedure, and the evaluation card. The data are used to group the judges into homogenous panels of high and low sensitivity using the group mean and individual thresholds, and to fix the approximate threshold for each panel.

ii) Final evaluation of the sample through the use of a dilution series, importantly, an arithematic series with small increments in concentration around the approximate threshold, maintaining one 'jnd' – which has been normally found to be one-tenth of the threshold concentration – and obtaining 15-20 judgements with a minimum of 5 panelists and of 3 to 4 repeat tests.

iii) Judgements are decoded in terms of Scoville units (SU = reciprocal of dilutions) and the mean value is expressed as a range, in SU $\pm\sigma$.

iv) The panel is defined by its average threshold in SU $\pm\sigma$ for pure capsaicin, or for an oleoresin of known capsaicin content.

The definition of the panel is of great help, since the values given by one panel could be converted to that of another panel of similarly defined sensitivity through the ratio of their thresholds to the same reference, e.g. capsaicin. This makes comparison of data from different sources possible. The discrepancy between different panel values used to be the main point of dispute earlier.

It has been shown that when pungency is determined through the just outlined procedure, and the total capsaicinoids content has been obtained by a reliable method, there is a highly significant linear regression ($P < 0.001$). The coefficient multiplying the dependent variable in the regression equation will reflect the acuity of the panel.

Todd et al (10) in their endeavour to establish a gas-chromatographic method for estimation of individual capsaicinoids for correlating pungency of capsicum and its extracts, studied the problems of sensory determination of pungency of individual capsaicinoids. They started with the ASTA 21.0 procedure, using four concurrent panels and repetitive testing with different dilution levels of capsaicin, and discussion to improve panel understanding procedures. Pungency data were based on bite frequency vs dilution. Computer analysis of data for each panel member, each panel, and the average of the four panels showed that uncertainty in the threshold pungency values was as much as ± 25%. Though this was considered normal for the Scoville method, it was unacceptable for correlation work.

They attempted to improve the panel performance by two modifications in the ASTA 21.0 method. First, the test was done as a triangle test with a judgement on intensity (0 – no bite, 1-slight, 2-moderate and 3-strong bite) so that pungency levels greater than threshold levels could also be used to obtain the actual threshold by extrapolation. Besides, panelists being slightly insensitive, many tended to report false 'bites' when dilutions were past the extrapolated threshold value. The preferred panelists should have an approximately linear intensity rating response over a large dilution range while trailing off at the barely perceptible level. This would make the panel homogenous. The estimates by four panels gave an average threshold value of 16.2±2.6 million for capsaicin, but the variance of the method was still as high as ±16%. As a second modification, the dilutions close to the estimated threshold were varied by small increments, and tests carried out as above, employing a triangle test with intensity judgements. The data handling, however, was changed by assigning +1 when the odd sample was correctly identified and positive 'bite' reported, and -1 either when the odd sample identification was wrong, or when no bite was reported in the correctly identified odd sample. The totals of the data of the five panelists were taken as one data point for that particular dilution level. A computer was used to find the best fitting n^{th} degree polynomial through the assembled data points for different dilution levels, and the threshold pungency was determined. This gave values varying between 15.8 ± 0.6 to 16.8 ± 0.6 million for capsaicin by the four panels, and uncertainty was reduced to about ± 5%. Thus, the magnitude of concentration difference between samples in the series appears to be the main source of variation in pungency evaluation.

The threshold values obtained by this improved procedure for the different capsaicinoids and some synthetics are given in Table X. These values were used to calculate actual pungency of capsicum extract from the gas chromatographic determination of individual capsaicinoids, and a significant correlation with sensorily estimated pungency was found (See later, Figure 3).

Thus, the methods described above, both of which functionally require the same tasting facilities and time as the official ASTA method, are capable of yielding reliable values which are statistically acceptable and useful for correlation work. The method developed in our laboratory with the simpler, prescribed, dilution series has been successfully used for testing pungency of pepper and ginger also, and has been adopted as an official method for pungency determination by the Indian Standards Institution (14).

Specific Components for Pungency

The clearly identifiable and recognised pungent components in food use are those from the spices used as flavouring agents. Some mild pungency is also felt in some vegetables belonging to the Cruciferae family. As purified components, these vary by several orders of magnitude in their response intensity, and are also of very varied organic structures, ranging from the simpler volatile phenols, such as eugenol in cloves, and isothiocyanates in the vegetables, to non-volatile alkylamides and alkylketones in the pungent spices. Comprehensive reviews on the three spices, capsicum, pepper, and ginger, have appeared recently (15,16,17). Hence, the chemistry of the specific pungency stimuli in these spices, the methods chosen for the physico-chemical determination of these stimuli, and the correlation of these values with pungency will be briefly discussed.

Chillies (Capsicum Sp.)

The most important of the spices known for pungency are the chillies, the larger and less pungent variety, *Capsicum annum* L., the smaller, pungent variety *Capsicum frutescens* L. and the bird chilli a *Capsicum frutescens* variety. Apart from color, which is important in the case of paprika, a large variety, pungency is the most important quality attribute of the capsicums used in foods. The capsaicinoids have long been known to cause the pungency response.

Table I. Capsaicinoids and pungency of some world varieties of capsicum

Samples	Capsaicinoids, %		Pungency (SU)
	Determined	Calculated	Determined
Cayenne red pepper	0.2360	--	40,000
Red pepper	0.0588	--	10,000
Chilli	0.0058	--	900
Mombasa (Africa)	--	0.800	120,000
Uganda (Africa)	--	0.850	127,000
Mexican pequinos	--	0.260	40,000
Abyssinian	--	0.075	11,000
Bahamian (Bahamas)	0.5100	--	75,000
Santaka (Japan)	0.3000	--	55,000
Sannam (India)	0.3300	--	49,000
Bird Chilli (India)	0.3600	--	42,000

Data from (18,19,20). Pure capsaicin = 15 to 17 x 10^6 SU.

The average capsaicinoids content and pungency of the chilli varieties, as collected from literature (18,19,20), are given in Table I. It is clear that there is much variation in the capsaicinoids content. The corresponding pungency values seem to show proportionality. The capsaicinoids are a group of related compounds, vanillylamides of monocarboxylic acids, varying in length (C_8 to C_{11}) and unsaturated. Since the early twentieth century, a number of synthetic compounds, mostly from straight-chain, saturated acids have been synthesised as substitutes. Table II gives the natural capsaicinoids that have been identified (10,21, 22,23), and Table III, the synthetic capsaicinoids (10,24), along with some relevant properties. The composition of capsaicinoids of natural origin, is generally: capsaicin, 70%; dihydrocapsaicin,

20%; nordihydrocapsaicin, homocapsaicin, and homodihydrocapsaicin together forming about 10% (15,21). However, in recent years, improved methods of separation and analysis of many varieties of chillies from different growing regions have shown that large variations exist (23). The principal components, capsaicin and dihydrocapsaicin, are reported to account for only 60 to 70% of the capsaicinoids in many samples while the other three formed upto 30% (25).

Hence, in such samples, it is necessary to know the pungency of the individual capsaicinoids. Recent careful determinations of the pungency thresholds of individual capsaicinoids have shown that the two major components, capsaicin and dihydrocapsaicin, have the same pungency, while the minor related components and the synthetic compounds have only about half, or less, of their pungency (10,24) (Tables II and III).

Table II. Pungency stimulants of capsicum - the natural capsaicinoids

General formula:- $R-CO-NH-CH_2-C_6H_3(OCH_3)(OH)$

R	Name (Abbreviation)	Mol. wt.	Composition, % of total: Range	Composition, % of total: Average	Relative Pungency
$(CH_3)_2-CH-CH\overset{t}{=}CH-(CH_2)_4-$	Capsaicin, (C)	305	46-77	70	100
$(CH_3)_2-CH-(CH_2)_6-$	Dihydrocapsaicin, (DC)	307	21-40	20	100
$(CH_3)_2-CH-(CH_2)_5-$	Nordihydrocapsaicin, (NDC)	293	2-12	--	57
$(CH_3)_2-CH-CH\overset{t}{=}CH-(CH_2)_5-$	Homocapsaicin, (HC)	319	1-2	--	43
$(CH_3)_2-CH-(CH_2)_7-$	Homodihydrocapsaicin, (HDC)	321	0.6-2	--	50

Data from (10,22,23)

Table III. Synthetic capsaicinoids

General formula:- $R-CO-NH-CH_2-C_6H_3(OCH_3)(OH)$

R	Name (Abbreviation)	Mol. wt.	Relative Pungency*
$CH_3-(CH_2)_6-$	Vanillyl octanamide, (VO)	279	50
$CH_3-(CH_2)_7-$	Vanillyl pelargonamide, (VP)	293	57
$CH_3-(CH_2)_8-$	Vanillyl capramide, (VC)	307	28
$CH_3-(CH_2)_9-$	Vanillyl undecanamide, (VU)	321	21
$CH_2=CH-(CH_2)_8-$	Vanillyl undecenamide	319	--

* - Relative to natural capsaicin as 100. Data from (10,24)

With this as background, we could decide on methods of estimation of capsaicinoids in capsicum by instrumental methods. Many methods, utilising partition, column, paper, and thin-layer chromatographic separations, combined with colorimetry and gas chromatography, have been suggested for estimation of the capsaicinoids (26). All these earlier methods give estimates of the total capsaicinoids; but the recent gas chromatographic method after silylation is capable of separating and estimating the individual natural capsaicinoids and some of the synthetics (10).

For routine quality evaluation, it is considered sufficient to determine accurately the total capsaicinoids; the minor components generally amount to about 10%, and have 50 to 60% of the pungency response of the major capsaicinoids. For this purpose Salzar (26) compared a number of the earlier methods, and found the paper chromatographic method (27) to be simple, rapid, reproducible and accurate. The method is applicable directly to the oleoresin, and is based on the fact that, of the three major components of a total extract, the capsicum color would be adsorbed by the paper and along with the fat would be moved little by the buffered methanol solvent, while the capsaicinoids are clearly separated, moving to a high Rf position (Figure 1). The accuracy of the determination was also improved by the use of the more sensitive and specific Gibbs' reagent for the colorimetry.

However, the values by any of these methods measuring total capsaicinoids would give a low correlation with pungency, when there are high proportions of the C_8 and C_{10} capsaicinoids, or when there is gross adulteration with synthetics. The detection of adulteration by synthetics has been attempted since 1955, but all the methods were rather complex, and so, unsatisfactory for rapid and routine screening. Recently, Todd et al (28) have developed elegant, rapid, thin-layer chromatographic methods for separation of capsaicinoids and synthetics by utilizing the difference in solubility between the different capsaicinoids as the chain length increased, and by switching to partition chromatography in the reversed phase, with silver-ion complexing, to separate the saturated and unsaturated compounds. A separation of the five synthetics from the natural capsaicinoids as also the five natural capsaicinoids into three spots was achieved in the first direction.

Table IV. Capsaicinoids and related synthetic compounds - Thin layer chromatography, R_f values

Compounds	Reversed Phase			Polyamide Methanol (Ag^+)/ water/acetic acid 52:40:8 (v/v)
	Methanol/water 60:40 (v/v)		Methanol (Ag^+)/ water/acetic acid 52:40:8 (v/v)	
	As such	Brominated		
Vanillyl octanamide	0.70	0.70	---	--
Vanillyl pelargonamide	0.64	0.64	--	--
Capsaicin	0.52	0.65	0.39	0.64
Nordihydrocapsaicin	0.52	0.52	0.25	0.42
Vanillyl capramide	0.44	0.44	--	--
Dihydrocapsaicin	0.41	0.41	0.17	0.32
Homocapsaicin	0.41	0.63	0.30	0.57
Vanillyl undecenamide	0.36	0.25	--	--
Homodihydrocapsaicin	0.31	0.31	0.10	0.26
Vanillyl undecanamide	0.20	0.20	--	--

Data from (28).

Bromination of the separated spots by exposure to bromine vapor in a tank, and further development in the second direction achieved the separation of the bunched unsaturated from the saturated natural components with one methylene group less. The natural, saturated and unsaturated capsaicinoids could be clearly separated in a single development, and more rapidly by the use of reversed phase, or polyamide, plates, and using developing solvents with silver ion for complexing. The separation possibilities are clear from the R_f values for the different systems given in Table IV. These methods are applicable to the extracts or diluted oleoresins directly, not requiring any pre-separation; hence, they are useful for rapid screening in the field for detection of adulteration, or for selection of improved strains in agronomic and genetic work.

The estimation of individual components and different synthetics was effectively achieved by gas chromatography after simple silylation (10). Figure 2 shows the separation obtained, clearly away from other volatiles in a capsicum extract. The individual capsaicinoids were quantitated in relation to vanillyl octanamide as the internal standard chosen, because it eluted just prior to the capsaicinoids and the response factors of the capsaicinoids with reference to this standard were close to unity. The standardised conditions were as follows: Silylation with N,O-bis (trimethylsilyl)-trifluoroacetamide in tetrahydrofuran gave clear rapid reaction at room temperature. The silylated extracts were injected directly on to a stainless steel column of 2 m x 3 mm, filled with 3% SE-30 on Chromosorb-GHP, (100-120 mesh); the column temperature was programmed from 170° to 215°C at 4°/min. and held at 215°C for 10 minutes. The injection port temperature was important for rapid volatilisation of all the components without decomposition, and was fixed at 200°C. The flame ionisation detector temperature was kept at 250°C, and nitrogen flow at 20 ml/min. The percentages of the individual capsaicinoids were calculated from the areas of the peaks, the response factors, and the weight and area of the reference compound.

Pungencies of samples were calculated bymultiplying the percentages of individual capsaicinoids from gas chromatography with the corresponding threshold pungency, and the total was expressed in Scoville units. Figure 3 illustrates the close relation between the pungency calculated from the gas chromatographic composition and the directly (sensorily) estimated value of pungency. A correlation coefficient greater than 0.95 was obtained for pungency values varying from 50 M to 2.0 MM.

It is, therefore, clearly possible to evaluate by instrumental methods the quantity of synthetics present as adulterants in a sample of capsicum extract or oleoresin, to determine individual capsaicinoids or total capsaicinoids, and to predict the pungency of the capsicum preparations.

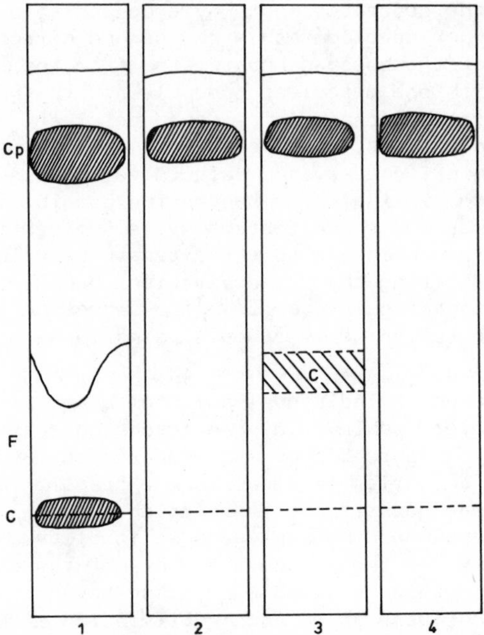

Figure 1. Tracings of separation of total capsaicinoids of capsicum extracts and reference pure compounds by paper chromatography: Spot (1) red capsicum; Spot (2) natural capsaicinoids pure; Spot (3) green capsicum fresh; and Spot (4) synthetic capsaicin; (C) color; (F) fat, and (C_p), capsaicinoids/capsaicin ((27) and unpublished).

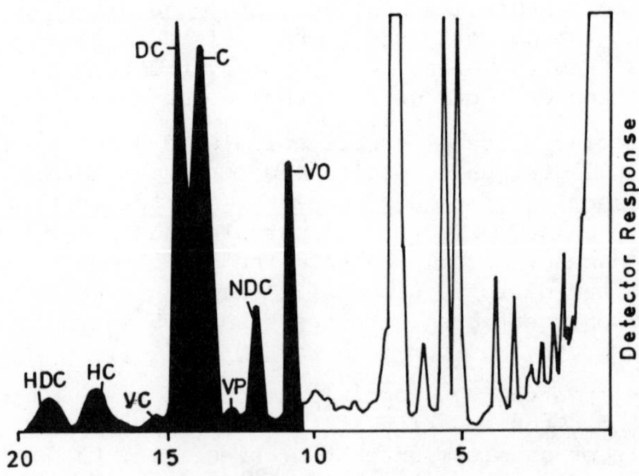

Figure 2. GC pattern of silylated capsicum extract (See text for GC conditions and Tables II and III for abbreviations) (10)

Pepper (*Piper nigrum* Linn.)

Pungency is the quality which has made pepper world famous as the King of Spices; it is given an even greater weightage than aroma when determining the overall quality of pepper.

The chemistry of the pungent compounds of pepper has been under study since the isolation of piperine in 1820. Piperine was shown to be a piperidide of piperic acid, and had the *trans*, *trans* configuration. The three other possible isomers were soon postulated, and they were named isopiperine (*cis*,*trans*), isochavicine (*trans*, *cis*) and chavicine (*cis*,*cis*) even before they were synthesized. The assignment of the isomeric configuration starts from the amide end.

There was much controversy about the identity of the highly pungent compound in the mother liquor, after crystallisation of piperine from pepper extracts. Bucheim's claims of this compound being the *cis*-*cis* isomer and being more pungent than piperine were believed for a long time, even though it was clearly shown that piperine was not pungent in the crystalline state, but, was extremely pungent as a solution in alcohol. The early studies are reported in detail in a recent review which also gives a comprehensive bibliography (16).

Grewe *et al* (29) synthesised the four possible isomeric acids by stereospecific routes, as also the corresponding piperidides, and determined their spectral and other physical characteristics. This cleared the confusion in earlier works on the supposed isolation of chavicinic acids from pepper. They also determined their pungency and showed that, excepting piperine, 5(3,4-dioxymethylene phenyl)-2-*trans*,4-*trans*-pentadienoic acid, all other isomers are poorly pungent. They concluded that neither chavicine nor the other two, *cis*-*trans* or *trans*-*cis*, isomers exist in natural pepper. They also showed that most of the isomers were unstable to light and were easily converted into the *trans*,*trans* piperine.

DeCleyn and Verzele (30,31,32) isolated the four possible isomers from piperinic acid irradiated in an ultra-violet reactor, by counter-current distribution; and also the piperidides obtained synthetically from the treated piperinic acid by high pressure liquid chromatography (Figure 4). The structures of the isomers were derived mainly from NMR data. Their pungency was recorded, but possibly not by rigorous methods (31,32). The results confirmed the observation of Grewe *et al* (29) that piperine is the pungent principle of pepper, and that other isomers have little 'taste'. The data on the properties of the isomers are collected in Table V.

DeCleyn and Verzele had originally (30) held the view that the loss of pungency in pepper on storage is possibly due to light-induced isomerisation of a part of the piperine to the other isomers which are not pungent. But, in their latest work (33), they find little evidence of the presence of isomers in pepper samples

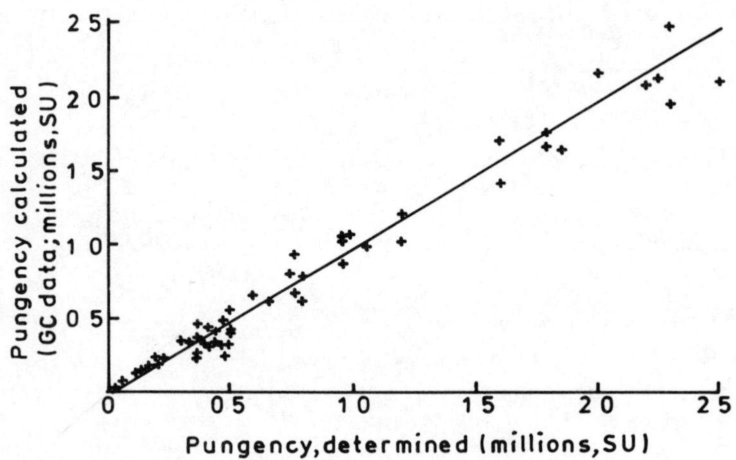

Journal of Food Science

Figure 3. Correlation of pungency values (in millions, SU) estimated and calculated from GC determination of individual capsaicinoids (10)

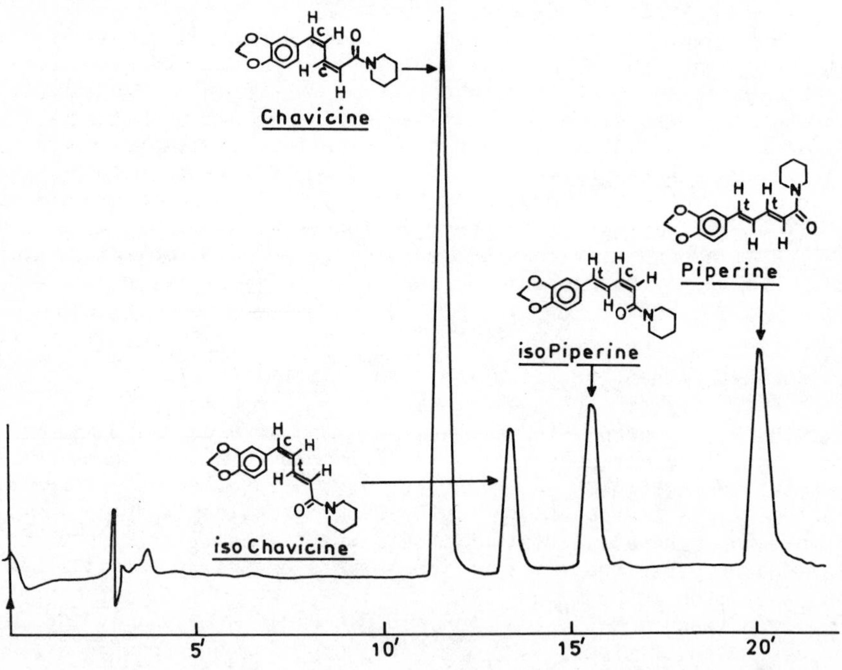

Figure 4. HPLC separation of photostationary state mixture of UV-irradiated piperine (33)

Table V. Piperine and isomers and their properties

General Formula:-

5 4 3 2 1

H_2C (O, O) – –CH=CH–CH=CH–CO–N

Common name	Isomeric structure	m.p. °C	Absorption data		Pungency (Relative)
			max. nm	Molecular	
Piperine	2-t,4-t	128	343	32,000-37,000	+++++
Iso-piperine	2-c,4-t	110	332	21,800	+
Iso-chavicine	2-t,4-c	89	333	16,300	+
Chavicine	2-c,4-c	Oily	318	16,200	++

Data from (29).

stored over 10 to 15 years, and conclude that the photoreaction does not occur in the solid state. Our work (34) on the quality of different cultivars of pepper grown and stored in India has also shown that, when stored for over 5 years as wholes, there is hardly any reduction of their piperine content as measured by the characteristic 342 nm absorption of the extracts. Other detailed analyses in our laboratory, by thinlayer and gas chromatography of dilute pepper extracts, and of piperine solutions exposed to light, have shown that, at high dilutions, piperine and its isomers are very labile and result in constantly changing mixtures of isomers. Piperine in whole pepper, or in concentrated oleoresins, protected from exposure to strong light, could, however, be considered to be stable as validated by careful pungency tests (34)(Figure 5).

Certain analogs and homologs of piperine have been reported in pepper as minor constituents. The earliest of this was piperettine, the heptenoic analog of piperine with three double bonds, giving it the characteristic absorption maximum at 360 nm. This compound shows substantial absorption at 342 nm, the absorption maximum of piperine, and thus, would enhance the piperine value. No separation methods have been available but only an estimation based on measurements at 342 nm and 360 nm of an extract, and use of a simultaneous equation. However, piperttine which, according to differential absorption measurements, makes up 5 to 15% of the piperine content (35), has not been referred to in any of the many subsequent investigations on pepper with modern techniques. Its contribution to pungency has also not been determined. Being present in substantial amounts in some of the samples, it could lower the established pungency equivalent based on piperine content, in case pure piperettine is shown to have no pungency.

We have recently been able to show (34) the presence of a minor component in extracts of many cultivars of pepper grown in India and Ceylon by the reversed phase, argentation thin-layer chromatography of the isolated total pungent components. This component has the typical 360 nm maximum attributed to piperettine and has shown very little pungency. We are still working on further isolation and purification for study by mass and nuclear magnetic resonance spectroscopy.

A pyrollidine analog of piperine, piperylin (syn. pyrroperine) has been reported by both Grewe et al (29) and Mori et al (36) in recent years. This possibly occurs in the order of 2-4 percent of piperine in pepper. Its presence in another Piper species has been reported, and it was synthesized much earlier (see (7). Systematic pungency evaluation has not been done, for it is reported to be equally pungent to piperine (29), as also poorly pungent (36). This analog has similar absorption characteristics as piperine, and would be measured in the method based on 342 nm absorption. Being present in such a small amount, and having some pungency, it may not contribute much to the error in the estimation of the total pungency simulants in pepper.

A monoenoic analog of piperine, piperanine has recently been reported by Traxler (37) in Malabar pepper. This compound was isolated after many chromatography steps, and purified for determination of its properties, and also synthesized. It was reported to be half as pungent (details are not clear) as piperine. Piperanine, however, is present in such a small proportion as not to affect the subjective pungency evaluation; and not having any absorption at 342nm, it will not be determined as piperine. Table VI gives structural details and physical properties of these analogs of piperine in pepper.

On tasting the narrow, horizontal segments of the area from the start to the solvent front in many thin-layer separations of pepper extracts, we have noticed that there is a component always coming above the piperine spot which has a tingling sensation on the tongue. This substance has not been identified, but could consist of isobutylamide-like compounds reported from other Piper species by Atal et al (38).

Colorimetry, hydrolysis-distillation, and spectrophotometric measurement at 342 nm have all been proposed for estimation of piperine and related components, the pungency stimulants in pepper. These methods have been critically compared in two recent reviews (16,26).

Table VI. Piperine analogs in pepper

Common name	Alkyl chain length	Isomeric structure	Mol. wt.	m.p. °C	max. nm	Pungency
Piperine	C_5	2-t,4-t	285	128	342	+++++
Piperanine	C_5	2-t	287	77	283; 233	++
Piperylin	C_5	2-t,4-t	271	142	345; 309; 261; 245	+++++ or +
Piperettine	C_7	2,4,6-enoic	311	146	364	?
Piperolein	C_7	6-t	315	Oily	-	-
Piperolein	C_9	8-t	343	Oily	269; 260; 214	-

Data from (29,36,37).

Our work over the years showed (39), that the earliest method, absorption at 342 nm is easily the best(being applicable directly to extracts) and most accurate, provided that reasonable precautions are taken to protect the extracts from undue exposure to light, and that benzene is used for the large dilutions necessary for the spectral measurement. The molecular absorption is so high that the estimation can be comfortably done on single corns, as would be necessary in work with rare samples, or in genetic work. The hydrolysis-distillation method and direct 342 nm absorption on extracts show a linear relationship with sensory pungency expressed in Scoville units, but the slopes differ, being lower than those for pure piperine - Scoville units (Figure 6).

The only interfering components (which are likely to lessen the correlation of the estimate of the active constituents by the 342 nm absorption to pungency) are the varying contents of piperettine (whose contribution to pungency appears negligible); the content of the still minor amounts of piperylin or pyrroperine (whose relative contribution to pungency has not been unequivocally established); any absorptions at 342 nm contributed by non-pungent color; and the tingling compound. We have shown that by a simple thin-layer run with hexane + ethyl acetate (90:10, v/v), the colored compound and the tingling sensation compound are separated from the total piperine and closely related compounds. The active

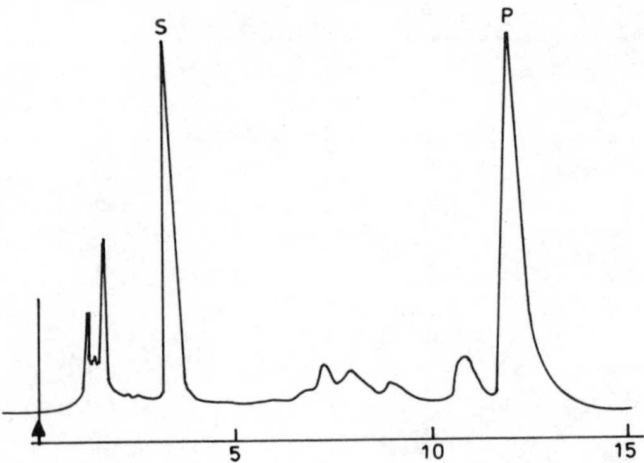

Figure 5. HPLC of extract of long stored pepper: (S) internal standard; (P) piperine (33)

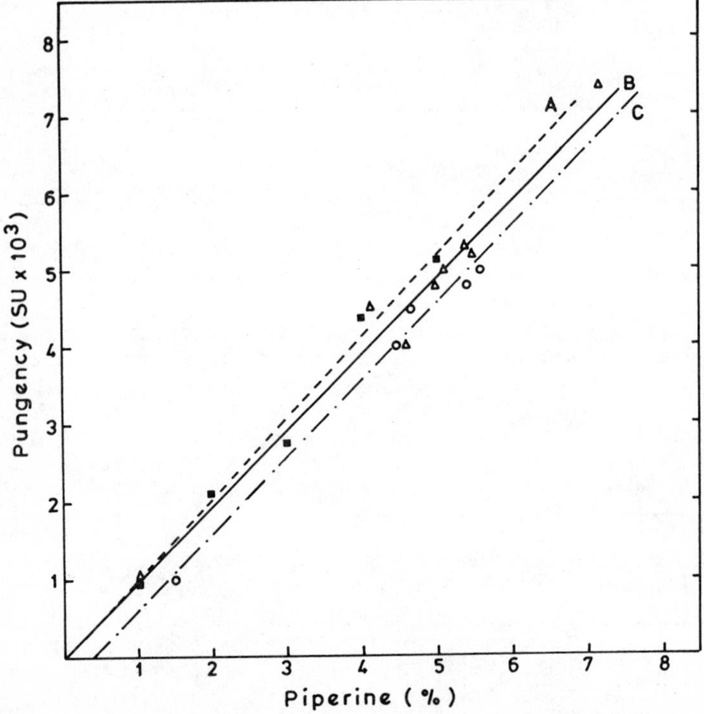

Figure 6. Correlation of piperine content and pungency. Curves: (A) pure piperine; (B) piperine by direct 342 nm absorption of pepper extracts; (C) piperine by Labruyere's hydrolysis–distillation method (39)

compounds can then be extracted with ether and measured at 342 nm.

Thus, the estimation of piperine (along with the minor amounts of piperettine and piperylin) obtained by a measure of the absorption at 342 nm of the extract has been shown to be reasonably accurate (Table VII), and predictive of the real pungency of pepper samples for routine quality control. The estimation of the piperine by 342 nm absorption has been approved by the Indian Standards Institution and by the American Spice Trade Association. The latter, however, still keep, as a recommended method, a colorimetric method which has been repeatedly shown to give very variable results.

Table VII. Piperine content of world varieties of pepper and their pungency

Sample/(Country)	Per cent dry wt[a]	Pungency[b] SU × 10^3
Black		
Tellichery (India)	5.2	5.3 ± 0.3
Malabar (India)	5.1	5.0 ± 0.2
Panniyur (India)	5.3	5.8 ± 0.1
Ceylon (Sri Lanka)	6.7	6.5 ± 0.3
Light Pepper (India)	5.2	5.3 ± 0.2
Pin head (India)	1.0	1.0 ± 0.1
Lampong	5.3	----
Sarawak	5.7	----
Sumatra	5.1	----
White		
Muntok	5.1	----
Sarawak	5.6	----
Ceylon	8.5	----
India	4.5	5.0 ± 0.2

a - Determined by absorption at 342 nm;
b - Determined according to (13,14).

Data from (35,39)

Ginger (*Zingiber officinale* Roscoe)

Ginger is another spice which contributes both aroma and mild pungency to foods. Possibly because the chemistry of the pungent components was not well established till the early seventies, and no simple quantitative method was available, the specifications and standards refer only to percent volatile oil as an index of quality, with no reference to pungent compounds.

Early Japanese work, which has been critically studied and expanded by Connell and colleagues (40,41), has clearly shown that pungency in fresh ginger is due to the homologous group of phenylalkyl ketones, the gingerols. The dominant member of this group is the (6) -gingerol, the prefix (6)- indicating the hexanal that would be obtained by alkali degradation. These gingerols have a labile β-hydroxy-keto grouping, and have been shown to be susceptible to pH-dependent, thermal dehydration to the corresponding β-unsaturated compounds, the shogaols. Under more drastic conditions, the latter degrade further to give the simpler abbreviated ketone, zingerone, and the corresponding aldehydes. These relationships are shown in Figure 7.

The presence of the higher homologs in ginger was established by Connell and Sutherland (41), by alkaline degradation and identification of the released aldehydes. Connell (40) showed that the gingerols and shogaols are clearly separated by thin layer chromatography on silica-gel, but did not develop a quantitative method. With our interest in the pungency of spices, we showed (42) - by incorporating a taste testing step on a co-chromatographed spot in thin-layer chromatograms of ginger extracts - that, in the rather elongated spots obtained in Connells' separations, only the lower portions in both the gingerol and shogaol areas showed pungency, while the other portions were not pungent or had very little pungency. We have also, recently, improved on the separation of the component gingerols and shogaols by adopting the wedging technique, and have established conditions for estimating the pungent and poorly pungent components with the help of the Folin-Ciocolteau reagent (43). Figure 8 shows the clear separations of the individual homologs, as compared to the earlier separations in groups.

Following preliminary observations by Connell and McLachlan (44), on gas chromatographic behaviour of gingerols and shogaols, we have just established improved conditions for the study, using the purified components obtained from thinlayer chromatograms. During programmed-temperature gas chromatography (60° to 260°C at 8°/min, injection port at 250°C), a part of the isolated total gingerols underwent pyrolytic breakdown to the aldehydes and zingerone, and the rest was dehydrated to the corresponding shogaols, while the isolated shogaols eluted out without any change. Figure 9 gives a composite picture of the separations for total and the thinlayer separated components. We have clearly shown (45), through the identity of the aldehydes and shogaols, that the (6)- gingerol and (6)- shogaol are the principal components responsible for the pungency, while the higher homologs have little or no pungency.

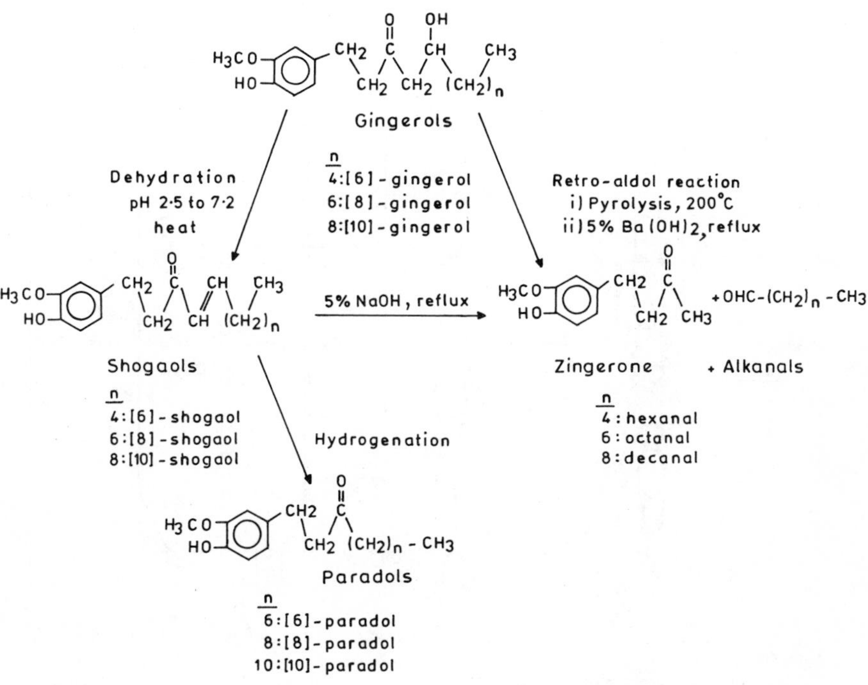

Figure 7. Relationship of gingerol and derived compounds in ginger extracts (45)

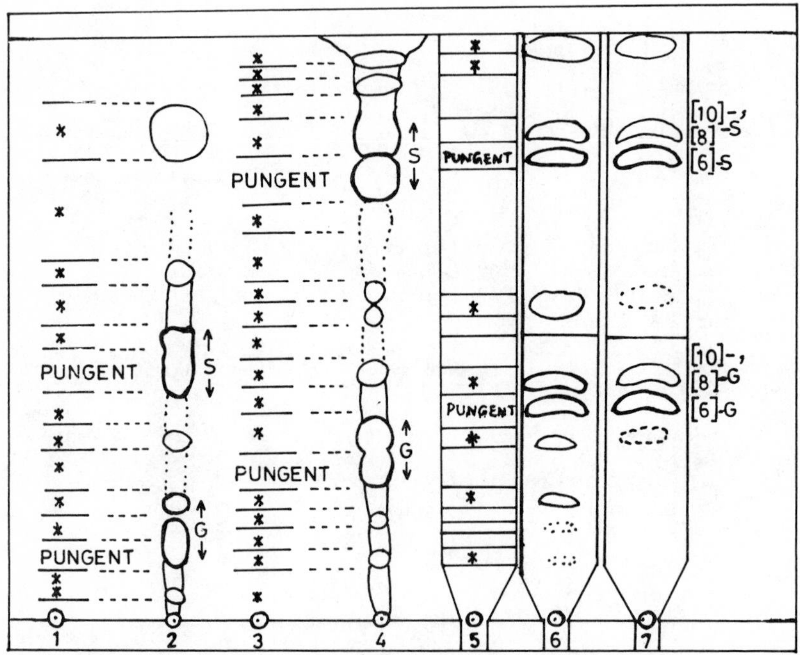

Figure 8. Tracings of TLC separations of the homologs of gingerol and shogaol. Conditions of separation for Spots 1 and 2 according to (40); 3 and 4 according to (42); 5, 6, and 7 according to (43). Spots 2, 4, and 6 sprayed with phenol reagent; 7 sprayed with 2,4 DNPH reagent; spots 1, 3, and 5 cochromatographed spots, taste tested.

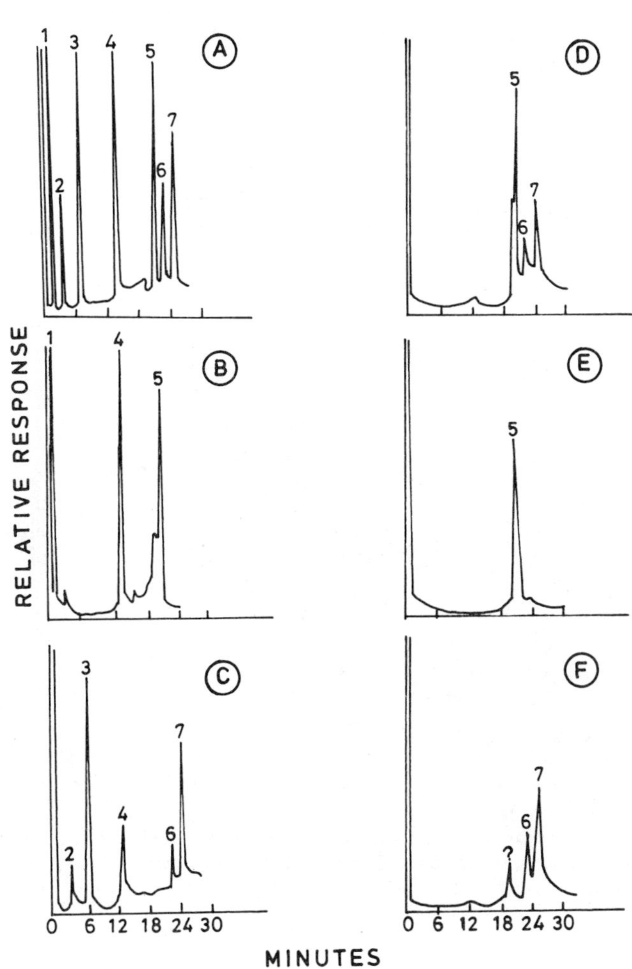

Journal of Food Quality

Figure 9. GC analysis and identity of purified gingerol and shogaol homologs (See text for GC conditions) (A) total, (B) pungent, and (C) nonpungent gingerols; (D) total, (E) pungent, and (F) nonpungent shogaols. Identity of peaks: (1) hexanal; (2) octanal; (3) decanal; (4) zingerone; (5) 6-shogaol; (6) 8-shogaol; and (7) 10 shogaol (45)

The total extracts or diluted oleoresins could also be analysed under these gas chromatographic conditions. While the aldehydes emerging early are mixed up with the volatile oil components, the zingerone and the shogaol peaks elute separately and later. The ratio of the zingerone to shogaol provides a rapid screening technique for finding fresh and stored oleoresins. Fresh oleoresins , containing essentially gingerols, show a high zingerone peak; stored oleoresins, in which the gingerols had slowly changed to shogaols, showed a low zingerone and high shogaol peaks (17).

In the direct analysis of ginger extracts by gas chromatography, certain minor peaks are seen in the shogaols area. These are possibly from the minor amounts of lower and higher homologs and related acetates of gingerol recently reported (45a). Their contribution to pungency is not known, but in view of the relation of pungency to chain length established in the case of the capsaicin (10,24) and paradol (46) series, minor amounts of these lower and higher homologs would not be expected to contribute significantly to pungency.

It had been assumed by Connell (quoting Kulka) (40), that the change, gingerols to shogaols to zingerone, resulted in progressive reduction in pungency, probably based on the fact that old, long-stored oleoresins are of definitely poor sensory quality. We have carefully studied (47), by the standardized pungency evaluation method (13), many ginger oleoresin samples obtained from different varieties, and under varying conditions of processing and storage. These studies have shown that stored samples of oleoresins actually have more pungency; and this led to the finding that (6)-shogaol is about twice as pungent as (6)-gingerol. Only by assuming this ratio of pungency contribution, have we found a good linear relation between the chemical estimations of (6)-gingerol and (6)-shogaol in the samples and the subjective pungency (47 and Table VIII). A multiple linear regression worked out with values obtained from over thirty samples of varying gingerols and shogaols composition has shown that 86% of the variation is accounted for.

Zingerone has not been found by us either in fresh oleoresin samples, or in samples stored over some years; it will probably be found only under high thermal abuse of samples. A sample of zingerone obtained by alkali hydrolysis of gingerol and a synthetic sample have been tested, and found to be only mildly pungent.

Thus, the pungent principle in ginger has been shown to have a complicated composition, in that it consists of two related components which vary in their proportion, depending on processing and storage. However, the active components, (6)-gingerol and (6)-shogaol can be reliably determined by thinlayer separation and colorimetry, and the pungency of a sample can be predicted by

Table VIII. - Ginger oleoresins - Active components and pungency

Samples	% Active components		Pungency SU x 10^3	
	(6)-G	(6)-S	Est.	Cal*
Rio-de-Janeiro[a]	29.6	1.4	28.0	24.3
China[a]	22.2	1.8	17.0	19.0
Sierra Leone[a]	31.0	2.5	30.0	26.9
Ernad Manjeri[a]	21.7	2.0	22.4	19.2
Commercial, dry ginger[a]	9.2	13.6	26.0	27.3
Commercial, dry lime treated[a]	20.3	4.3	21.3	21.7
Dry ginger[b]	13.3	10.2	25.3	25.3
Dry ginger, stored[b]	18.9	10.9	27.5	30.5
Green ginger, stored[b]	25.2	10.9	38.1	35.2
Jamaican, stored[b]	8.8	12.6	28.7	25.5

a- Varieties grown in India; b- commercial samples

*- calculated values by using Scoville values of 150 x 10^3 for (6)-shogaol and 75 x 10^3 for (6)-gingerol.

Data from (47).

use of the Scoville values obtained for pure (6)-gingerol and (6)-shogaol.

Other spices

Other spices - e.g. mustard, cloves - are widely used as food additives to enhance palatability. Both are characterized by volatile components which produce two responses: pungency and characteristic aroma. The simple phenol, eugenol, is the active component in cloves; various isothiocyanates, allyl- in black, allyl- and 3-butenyl- in brown, characterize the mustards. The non-volatile p-hydroxybenzyl isothiocyanate present in white mustard shows only pungency. Definitive methods of estimation of these components are well established as quality parameters of these spices (48 Table IX).

Pungency in vegetables

Certain vegetables are also examples of foods with volatile components which stimulate both aroma and pungency. These are all sulphur compounds, present as precursors in the whole vegetable and are converted into the active constituents only by action of enzymes released when the structure is destroyed by cooking or grinding.

There is considerable variation in the pungency of the vegetables depending on the extent of processing they have undergone, and the completion of enzymic action, and the continuing chemical reactions. While the isolated compounds are extremely active as irritants, and often lachrymatory, the concentrations in the vegetables are only at the level at which they endow a desirable sensory reaction of mild pungency, or else they are modified by other reaction products such as nitriles found in water cress (50).

Horseradish (Armoracia lapathiofolia Gilib.) is the typical example of the vegetable class, but it is used more as a flavoring additive like mustard. Horseradish is peculiar in having in its volatiles relatively large concentrations of virtually only isothiocyanates. Allyl isothiocyanate, the principal stimulant of mustard which is known to stimulate pungent and lachrymatory responses, is also the most important component of horseradish flavor giving it the characteristic aroma and pungency. The small amount of allyl thiocyanate which is also present has only a garlicky aroma, but neither pungency nor any lachrymatory action. Another major isothiocyanate, the 2-phenylethyl isothiocyanate is reported to produce only a tingling sensation, but not the pungency nor the lachrymatory responses; the 2-butyl, and 4-pentenyl-isothiocyanates contribute only a distinctive green andacrid aroma (51).

The vegetable radish (Raphanus sativus L.) shows mild pungency attributed to 4-methylthio-3-butenyl-isothiocyanate. Other vegetables of the Brassica and Allium species, too, are reported to contain small amounts of these isothiocyanates, but they are possibly diluted and altered during cooking, and so have not been reported as exhibiting sensory pungency in foods.

The bulbous vegetables of the Allium group all have a characteristic aroma attributed to thiosulfonates and disulfides. These, like the isothiocyanates, are produced from precursors during cutting and cooking. While in the raw state, some members of this family - such as garlic and onion - exhibit on crushing

Table IX. Isothiocyanates and their flavor quality

General formula:- R-N=S

R		Source	Flavour Quality
CH_2=CH-CH_2-	allyl	Horse radish; mustard (black)	Pungency; Lachrymatory
CH_2=CH-CH_2-CH_2-	3-butenyl	Mustard (brown); small amounts in Brassica sp.	Pungency; aroma
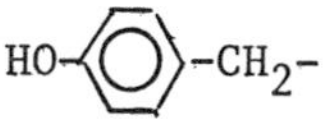 HO-C_6H_4-CH_2-	p-hydroxy-benzyl	Mustard (white)	Pungency; no aroma
CH_3-S-CH=CH-CH_2-CH_2-	4-methyl thio-3-butenyl	Radish	Pungency; sulphury aroma
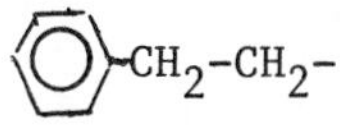 C_6H_5-CH_2-CH_2-	2-phenyl ethyl	Horse radish; water cress; turnip	Strong aroma of water cress; tingling sensation
CH_3-CH_2-CH-CH_3 (with bond from CH)	2-butyl	Horse radish	Acrid; leaf green aroma
CH_2=CH-CH_2-CH_2-CH_2-	4-pentenyl	Horse radish	Acrid; fragrant leaf

Data from (51)

Typical reactions are shown by the sequence below (49).

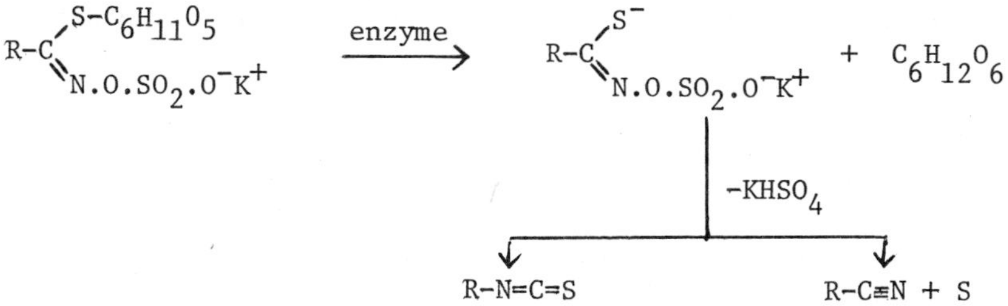

$$R\text{-}C\begin{cases} S\text{-}C_6H_{11}O_5 \\ N.O.SO_2.O^-K^+ \end{cases} \xrightarrow{\text{enzyme}} R\text{-}C\begin{cases} S^- \\ N.O.SO_2.O^-K^+ \end{cases} + C_6H_{12}O_6$$

$$\xrightarrow{-KHSO_4} R\text{-}N{=}C{=}S \quad \text{or} \quad R\text{-}C{\equiv}N + S$$

where R = allyl-; 3-butenyl, etc.; enzyme - glucosinolase.

in the mouth, a distinct action on the tongue. This property is not seen in cooked or roasted preparations.

The presence of a lachrymatory compound in freshly crushed onion has been known for a long time. This lachrymatory factor has been identified as 1-propenyl sulfenic acid (or isomerized to the more stable thiopropanal S-oxide), formed from the principal flavor precursor, _trans_ (+)-S-(1-propenyl)-L-cysteine sulfoxide. Other alkyl sulfenic acids are reported not lachrymatory, but are described as having 'biting astringency' or 'tongue tearing sensations' in common with the 1-propenyl sulfenic acid. These primary products being less stable, change rapidly to the respective thiosulfinates and are further converted to thiosulphonates and disulfides, the aroma compounds of the _Allium_ spices (52).

Structure and Pungency

The structure-response relationship in flavour studies is acquiring much importance in recent times with many laboratories and many disciplines becoming interested in the study. The properties of the chemical stimuli, the physiological response by the biological system, and the psychological interpretation and use of information by the higher centres in the human organism are all involved in these studies. We have learned a lot about the physicochemical properties of the flavor components of foods, but the details of the chemoreception, the receptor morphology and physiology, the interaction with the stimuli, the mechanism of perception are all still largely unknown. We can study the structure-response relationship as yet, only by the physicochemical parameters of the stimuli and the verbal response of the human instruments, the panelists. The response is the result of detection, quality recognition, intensity judgements, semantic capability, and preference (like/dislike); hence the necessity for careful selection and training of panelists to work in these correlative studies (4,7).

The physico-chemical parameters of the chemical stimuli which have been shown to have relevance and to be interrelated to the sensory response it elicits as specific odor or taste, are the factors controlling concentration at the receptor areas (solubility, hydrophilicity, lipophilicity, volatility, and partition coefficients), molecular features (size, shape, stereochemical and chirality factors and functional groups), and electronic features (polarity and dipoles) controlling positioning and contact at receptor surfaces (53). Many of these physico-chemical data are not available for many of the chemical stimulants, and till they are gathered, structure-response studies will be much restricted. The effects of interactions of the above parameters appear to a larger degree in the perception of odor, the dimensions of which are many and complex; viz. nuances, composite quality, multiple quality of single stimulus, besides intensity, and duration of sensation. In the case of taste, the informational complexity is much less - essentially, only a few primary types and their intensity - and pungency, like the four well-known tastes, varies only in intensity.

Against this background we will review the available data on the structures of natural compounds from spices established as stimulants for pungency, and on their natural and synthetic analogs.

A systematic study of structural variation in size and functional groups and their effect on pungency was attempted in the capsaicin, piperin and zingerone type compounds by different groups, early in this century. Unfortunately, the pungency determinations were not done by uniform methods and the results were often expressed in general literary terms; and, if a semi-quantitative scale was at all used, details are often lacking. In view of our present understanding of the problems in evaluation of pungency (described earlier in this review), the earlier statements and values on pungency should be considered relative and qualitative. Newman (54) had exhaustively and critically reviewed the available earlier literature, listing some 160 compounds related to capsaicin and piperine, and Provatoroff later summarized the work on the pungent compounds of ginger and synthetics related to zingerone (see 17). Since then, interest in structure and pungency has revived only during the last few years, in connection with careful estimation of the natural pungent compounds of capsicum, pepper, and ginger, and of some of their synthetic analogs. These have been discussed in some detail in recent reviews (16,17) and will be only summarized here.

The three groups of natural compounds which are pungency stimulants, the capsaicinoids, piperine, and gingerols, have some common structural features: viz., an aromatic ring, and an alkyl sidechain with a carbonyl function; while other features, such as the acylamide link, are common to capsaicinoids and piperine, and polar ends and vanillyl groups are common to capsaicinoids and gingerols. The length of the alkyl side-chain, the positioning of the amide function near the polar aromatic end, the nature of the groupings at the alkyl end, and unsaturation in the alkyl chain are the structural variations in those compounds which have been found to affect the intensity of pungency within each group. Some of these factors have been shown to reduce, and some to enhance pungency in the synthetic compounds tested. The natural compounds have a combination of these structural features. An overall comparison of the pungency of the natural compounds yields some interesting generalizations. Table X gives a selection of the natural and related synthetic compounds, their structures, sources and pungency in Scoville units. More compounds have been mentioned and discussed in earlier reviews (16,17,54).

Table X. Structure and pungency of natural stimulants and related synthetics

Formula	Name	Source	Pungency $SU \times 10^5$
$V{-}N(H){-}C(=O){-}(CH_2)_4{-}C(H){=}C(H){-}CH(CH_3)_2$ (t)	Capsaicin	Capsicum	160
$V{-}N(H){-}C(=O){-}(CH_2)_6{-}CH(CH_3)_2$	Dihydrocapsaicin	Capsicum	160
$V{-}N(H){-}C(=O){-}(CH_2)_5{-}CH(CH_3)_2$	Nordihydro-capsaicin	Capsicum	91
$V{-}N(H){-}C(=O){-}(CH_2)_5{-}C(H){=}C(H){-}CH(CH_3)_2$ (t)	Homocapsaicin	Capsicum	86
$V{-}N(H){-}C(=O){-}(CH_2)_7{-}CH(CH_3)_2$	Homodihydro-capsaicin	Capsicum	86
$V{-}N(H){-}C(=O){-}(CH_2)_6{-}CH_3$	Vanillyl octanamide	Synthetic	80
$V{-}N(H){-}C(=O){-}(CH_2)_7{-}CH_3$	Vanillyl pelargonamide	Synthetic	92.5
$V{-}N(H){-}C(=O){-}(CH_2)_8{-}CH_3$	Vanillyl capramide	Synthetic	45
$V{-}N(H){-}C(=O){-}(CH_2)_9{-}CH_3$	Vanillyl undecanamide	Synthetic	35
$V{-}N(H){-}C(=O){-}CH{=}CH{-}CH{=}CH{-}M$	Vanillyl piperamide	Synthetic	15
$V{-}CH_2{-}C(=O){-}CH_2{-}C(OH)(H){-}(CH_2)_4{-}CH_3$	(6)-Gingerol	Ginger	0.8

(Continued)

Structure	Compound	Source	Pungency
$V{-}CH_2{-}\overset{O}{\overset{\|}{C}}{-}\overset{H}{C}{=}\underset{\underline{t}}{\overset{H}{C}}{-}(CH_2)_4{-}CH_3$	(6)-Shogaol	Ginger	1.5
$V{-}CH_2{-}\overset{O}{\overset{\|}{C}}{-}CH_3$	Zingerone	Synthetic; from Gingerol	0.5 to 0.3
$V{-}CH_2{-}\overset{O}{\overset{\|}{C}}{-}(CH_2)_6{-}CH_3$	(6)-Paradol or Dihydro-(6)-Shogaol	Synthetic; Grain of Paradise	1.0
$V{-}CH_2{-}\overset{O}{\overset{\|}{C}}{-}CH_2{-}\underset{H}{\overset{OH}{C}}{-}(CH_2)_n{-}CH_3$	n=6; (8)-Gingerol n=8; (10)-Gingerol	Ginger Ginger	<0.1 <0.1
$V{-}CH_2{-}\overset{O}{\overset{\|}{C}}{-}\overset{H}{C}{=}\underset{\underline{t}}{\overset{H}{C}}{-}(CH_2)_n{-}CH_3$	n=6; (8)-Shogaol n=8; (10)-Shogaol	Ginger Ginger	<0.1 <0.1
$P{-}\overset{O}{\overset{\|}{C}}{-}CH\overset{\underline{t}}{=}CH{-}CH\overset{\underline{t}}{=}CH{-}M$	Piperine	Pepper	1.0
$P{-}\overset{O}{\overset{\|}{C}}{-}CH\overset{\underline{t}}{=}CH{-}(CH_2)_2{-}M$	Dihydropiperine	Pepper	1.0?
$P{-}\overset{O}{\overset{\|}{C}}{-}(CH_2)_4{-}M$	Tetrahydro-piperine	Pepper	1.0?
$P{-}\overset{O}{\overset{\|}{C}}{-}CH\overset{\underline{c}}{=}CH{-}CH\overset{\underline{t}}{=}CH{-}M$	Isopiperine	Synthetic	0.0
$P{-}\overset{O}{\overset{\|}{C}}{-}CH\overset{\underline{t}}{=}CH{-}CH\overset{\underline{c}}{=}CH{-}M$	Isochavicine	Synthetic	0.0
$P{-}\overset{O}{\overset{\|}{C}}{-}CH\overset{\underline{c}}{=}CH{-}CH\overset{\underline{c}}{=}CH{-}M$	Chavicine	Synthetic	0.0
$P{-}\overset{O}{\overset{\|}{C}}{-}CH{=}CH{-}CH{=}CH{-}CH{=}CH{-}M$	Piperettine	Pepper	<0.1
$P{-}\overset{O}{\overset{\|}{C}}{-}CH\overset{\underline{t}}{=}CH{-}(CH_2)_2{-}M$	Piperanine (dihydro-piperine)	Pepper	0.5?
$P{-}\overset{O}{\overset{\|}{C}}{-}CH{=}CH{-}(CH_2)_5{-}CH_3$	2-nonenoic piperidide	Synthetic	1.0

(continued)

P-C(=O)-$(CH_2)_7$-CH_3	Nonanoic piperidide	Synthetic	1.0
P-C(=O)-CH=(t)CH-CH=(t)CH-C$_6$H$_5$	5-phenyl-2,4-pentadienoic piperidide	Synthetic	1.0?
P'-C(=O)-CH=(t)CH-CH=(t)CH-M	Piperylin	Pepper	1.0?
V-CH_2-N(H)-C(=O)-$(CH_2)_9$-CH_3	Methylene vanillyl-undecanamide	Synthetic	0
V-CH_2-C(=O)-CH(OH)-CH(OH)-$(CH_2)_4$-CH_3	Gingediol	Ginger	0 (bitter)
$(CH_3)_2$CH-CH_2-NH-C(=O)-CH=(t)CH-$(CH_2)_2$-CH=(c)CH-(CH=(t)CH)$_2$-CH_3	Shanshol	Japanese pepper	mildly pungent

V is 4-hydroxy-3-methoxyphenyl-CH_2 (H_3CO-, OH- ring -CH_2) ; P is piperidino (ring N-) ; P' is pyrrolidino (ring N-) ;

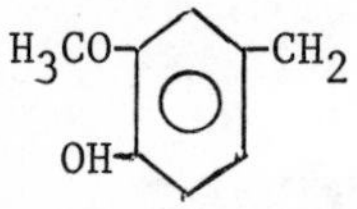

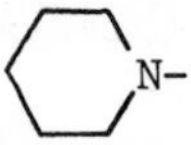

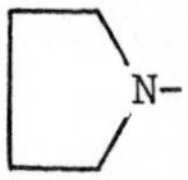

M is 3,4-methylenedioxyphenyl (ring with -O-CH_2-O-)

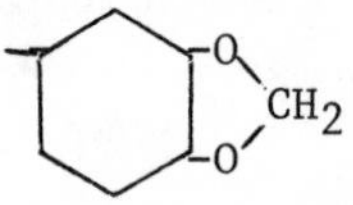

Data from different sources calculated to capsaicin as 160×10^5 SU. The values with a question (?) mark indicate variable values in literature.

Many compounds with fairly simple structures, such as phenyl methyl ketones, alkyl acylamides, exhibit pungency; but their intrinsic intensity varies by orders of magnitude; e.g. capsaicin, the vanillyl amide of 8-methyl-6-trans-nonenoic acid, elicits about 150 times as much pungency as piperin, the piperidide (3,4-dioxy-methylene phenyl)2-trans, 4-trans-pentadienoic acid, and gingerol, the 1-(4'-hydroxy-3-'-methoxy phenyl)-5-hydroxy decan-3-one. Within each series, the length of the sidechain required to produce maximum pungency varies within a narrow ranges; e.g., nonoyl vanillyl-amide, nonoyl piperidide are the most potent stimuli in their respective series, while the decan-3-one, (6)-gingerol is the most potent in the gingerol series.

The pungency potential is either abolished or markedly reduced when a hydroxyl is introduced in the side-chain; e.g., gingerol is reported to elicit bitterness and not pungency when the carbonyl group is reduced. Also, when the carbonyl group is separated by more than two intervening groups from the polar end, as when methylene vanillyl amine is coupled with the optimum C_9 aliphatic acid, the pungency potential is abolished.

Except for some stray observations, the introduction of unsaturation in the side chain has generally enhanced the pungency stimulation. However, in unsaturated compounds, the stereochemical factor assumes great significance. Recent work has shown that shogaol, which elicits more pungency than the corresponding gingerol (47) and the naturally occurring piperine, are both trans isomers; the other three isomers of piperine evoke little or no pungency (29,30). A phenyl group is not essential, for, a number of isobutylamides of aliphatic acids are known to elicit pungency, but when present has critical requirements. Placing of the aromatic ring in the middle of the alkyl chain, dioxymethylene substitution, or meta-hydroxy substitution of the aromatic ring, all result in lowering pungency response and are additive in their effects. The examples given have been selected essentially from the naturally occurring compounds, but the survey covers a large number of synthetic compounds also.

Though available pungency evaluation data on all compounds are not comparable, two features are striking:

i) a vanillylamide with an aliphatic acid of optimal length evoked a very high pungency response; and (ii) when there is unsaturation in the sidechain, pungency results only from the trans isomer.

The volatile isothiocyanates in mustard and horseradish elicit stronger responses both of pungency and lachrymation. This is possibly in line with earlier observations (2) that replacement of oxygen with sulfur or selenium, which belong to the same group VI of the periodic table, results in powerful and disagreeable reactions. Since other isothiocyanates, isolated from other vegetables, show only an acrid aroma, not identifiable with pungency, some other structural requirements - such as unsaturation and optimum chain length determining location of active centers - also seem to have a bearing on the sensation of pungency and its intrinsic intensity. Another structure containing a different functional group with a short chain, the

$$\overset{O}{\uparrow}\\ -SH$$

-SH (with O↑ above the S) or -CH=S=O in crushed onion is also associated with both lachrymatory and 'tongue tearing' (itchy?) sensations. Careful evaluations of these compounds by threshold tests and comparison with the recognised pungency stimuli are necessary.

The nature of the requirements with regard to shape, size, functional group, and stereospecificity of compounds that elicit a pungency response is, possibly, another argument in favor of considering pungency as a taste modality, and looking for the conformational structures contacting specific receptors.

It is of interest to note that the Shallenberger-Kier postulate of H.A-B features in the structures of compounds which produce

X

the sweet and bitter responses has further been applied by Beets (53) to structures of stimuli for sour and salt modalities. One may speculate whether the same postulate would be applicable, perhaps with a different optimal relative distance between the features to the structures of compounds which are established pungency stimuli.

Acknowledgement

The author thanks his colleagues Dr. K.G. Raghuveer and Mrs. Shanthi Narasimhan for the unpublished experimental work cited here, and Mr. K.M. Dastur for help in the preparation of this paper.

Summary

Pungency is defined as the gustatory sensory response to specific chemical stimuli found in certain spices and vegetables. It is distinguished from other sensations such as burning, irritation, lachrymation, and pain. Problems in evaluation of pungency are discussed, and standardized procedures that yield reproducible data which significantly correlate with the instrumental estimates of the corresponding stimuli are outlined. Specific pungency stimuli identified in foods are listed, and their chemistry, including reliable methods of estimation, has been briefly described. Attempts to correlate specific structural features of the stimuli with the chemoreceptory perception as pungency are reviewed.

Literature Cited

1. 'Sensory Quality and Ingestion' in "The Chemical Senses and Nutrition" M.R. Kare and O. Maller, Eds., Academic Press, New York, London, 1972.

2. Moncrieff, R.W., 'The Chemical Senses', Leonard Hill, London, 1967.

3. Singleton, V.L., and Noble, A.C., in 'Phenolic, Sulfur and Nitrogen Compounds in Food Flavors', G. Charalambous and I. Katz, Eds., American Chemical Society Symposium Series No.26, 1977, p.48.

4. Amerine M.A., Pangborn, R.M. and Roessler, E.B., "Glossary of Terms' in "Principles of Sensory Evaluation of Food", Academic Press, New York, London, 1965.

5. Boudreau, J.C., MBAA Tech. Quart., 1978, 15(2), 94.

6. "Sensory Quality Control - Practical Approaches in Food and Drink Production", Proc. Symposium, Symons, H.W., and Wren, J.J., Eds., Inst. Food Sci. Technol and Soc. Chem. Ind., London, 1977.

7. "Correlating Sensory and Objective Measurements", ASTM Symposium, STP 594, Americ. Soc. Testing Materials, Philadelphia, 1976.

8. "Optimizing Sensory Evaluation in Product Development", Symposium Inst. Food Technol., Food Technol., 1978, 32, 56-66.

9. Recommended procedures from Amer. Soc. Testing Materials, Committee E-18; Indian Standards Institution Committee AFDC-38.

10. Todd, Jr., P.H., Bensinger, M.G., and Biftu, T., J. Food Sci., 1977, 42(3), 660.

11. Scoville, W.L., J. Amer. Pharm. Assn., 1912, 1, 453.

12. American Spice Trade Association, "Official Analytical Methods", 1968.

13. Govindarajan, V.S., Shanti Narasimhan, and S. Dhanaraj, J. Food Sci. Technol., 1977, 14(1), 28.

14. Indian Standards Institution, Delhi, IS: 8104-1976; IS: 8105-1976.

15. Maga, J.A., Critical Rev. Food Sci., Nutr., 1975, 6, 177.

16. Govindarajan, V.S., Critical Rev. Food Sci. Nutr., 1977, 9(2), 115.

17. Govindarajan, V.S., Critical Rev. Food Sci. Nutr., in print.

18. Hartman, K.T., J. Food Sci., 1970, 35, 543.

19. Suzuki, J.I., Tausig, F., and Morse, R.E., Food Tech., 1957, 77, 100.

20. Mathew, A.G., Lewis, Y.S., Jagadishan, R., Nambudiri, E.S., and Krishnamurthy, N., Flavour Ind., 1971, 2(1), 23.

21. Bennett, D.J., and Kirby, G.W., J. Chem. Soc.,.1968, 442.

22. Kosuge, S., and Furuta, M., Agric. Biol. Chem., 1970, 34(2), 248.

23. Muller-Stock, A., Joshi, R.K., and Bucki, J., J. Chromatog., 1971, 63, 281.

24. Nelson, E.K., J. Amer. Chem. Soc., 1919, 41, 2121.

25. Boersma, J., Private communication, 1977.

26. Salzar, U.J., Internat. Flavors Food Addit., 1975, 6, 206.

27. Govindarajan, V.S., and Ananthakrishna, S.M., Flavour Ind., 1974, 5, 176.

28. Todd, Jr., P., Bensinger, M., and Biftu, T., J. Chromatog. Sci., 1975, 13, 577.

29. Grewe, R., Freist, W., Neumann, H. and Kersten, S., Chem. Ber., 1970, 103, 3752.

30. De Cleyn, R., and Verzele, M., Chromatographia, 1972, 5,346.

31. De Cleyn, R., and Verzele, M., Bull. Soc. Chim. Belges, 1972, 81, 529.

32. De Cleyn, R., and Verzele, M., Bull. Soc. Chim. Belges, 1975, 84, 435.

33. Verzele, M., Mussche, P., and Quereshi, S.A., under publication - Personal communication.

34. Govindarajan, V.S., Raghuveer, K.G. and Shanti Narasimhan, unpublished work.

35. Genest, C., Smith, D.M., and Chapman, D.G., J. Agric. Food Chem., 1963, 11, 508.

36. Mori, K., Yamamoto, Y., Tonori, K. and Komai, S.J., Food Sci. Technol., 1974, 21, 472 (in Japanese with English summary)

37. Traxler, J.T., J. Agric. Food Chem., 1971, 19, 1135.

38. Atal, C.K., Private communication.

39. Raghuveer, K.G., and Govindarajan, V.S., unpublished work.

40. Connell, D.W., Food Technol., Australia, 1969, 21, 570.

41. Connell, D.W. and Sutherland, M.D., Aust. J. Chem., 1969, 22, 1033.

42. Ananthakrishna, S.M. and Govindarajan, V.S., Lebensm. Wiss + Technol., 1974, 7, 220.

43. Bhagya and Govindarajan, V.S. under publication.

44. Connell, D.W. and McLachlan, R., J. Chromatog., 1972, 67, 29.

45. Raghuveer, K.G., and Govindarajan, V.S., J. Food Quality, 1979, 2(1), 41.

46. Locksley, H.D., Rainey, D.K., and Rohan, T.A., J. Chem. Soc. Perkins I, 1972, 23, 3001.

47. Shanti Narashimhan and Govindarajan, V.S., J. Food Technol., 1978, 13, 31.

48. Shankaranarayana, M.L., Raghavan, B., and Natarajan, C.P., Lebensm-Wiss + Technol., 1972, 5(6), 191.

49. Schwimmer, S. and Friedman, M., Flavour Ind., 1972, 3, 137.

50. MacLeod, A.J., and Islam, R., J. Sci. Food Agric., 1975, 26, 1545.

51. Gilbert, J., and Nursten, H.E., J. Sci. Food Agric., 1972, 23, 527.

52. Whitaker, J.R., Adv. Food Res., 1976, 22, 73.

53. Beets, M.G.J., "Structure-Activity Relationships in Human Chemoreception", Applied Science Publishers, 1978.

54. Newman, A.A., Chem. Prod., 1953, 16, 379, 467; 1954, 17, 14, 102.

Additional Reference

5a. Andres, K.H., and von During, M., in Handbook of Sensory Physiology, Vol. II, pp. 3-28; A. Iggo, Ed., Springer Verlag, Berlin, 1973.

5b. Jancso-Gabor, A. (Personal communication), 1979.

45a. Masada, Y., Inoue, T., Hoshimoto, K., Fujioka, M., and Shiraki, K., Internal. Cong. Food Sci. Technol. Abst., Madrid, Spain, 1974.

RECEIVED August 7, 1979.

4

Sweet and Bitter Compounds: Structure and Taste Relationship

H.-D. BELITZ, W. CHEN, H. JUGEL, R. TRELEANO, and H. WIESER

Institut für Lebensmittelchemie der Technischen Universität München and Deutsche Forschungsanstalt für Lebensmittelchemie München, West Germany

J. GASTEIGER and M. MARSILI

Organisch-Chemisches Institut der Technischen Universität München, Lichtenbergstrasse 4, D-8046 Garching, West Germany

Flavour impressions are conveyed by the senses of smell and taste. MOSKOWITZ (1) makes the following statement regarding the role played by these two senses:

"Smell is the most predominant in allowing us to form an overall impression, but the sense of taste plays an important part as well".

For various reasons sweet and bitter taste are important for many foodstuffs. Great interest is taken, for instance, in new sweetening agents, firstly because of the trend towards a diet containing fewer calories in many industrialized countries with their overweight populations, and secondly, because in various countries there have recently been renewed experiments and discussions, aimed at ascertaining how safe the sweetening agents saccharine and cyclamate really are. The search for new sweeteners is rendered more difficult by the fact that suitable compounds must not only be "safe", i.e. not harmful to the health, but must also satisfy various other criteria, such as that of sufficient solubility, stability even when exposed to extreme pH-values and temperatures. They must have sweet taste that is as pure and unadulterated as possible, with no by-or after-taste, and have a price which compares favorably with that of saccharose in terms of sweetening strength.

Bitter taste is an important component of a number of flavours of roasted foodstuffs. PICKENHAGEN et al. (2) have demonstrated, for example, that the bitter taste of cacao is caused by 2.5-dioxopiperazines, which evolve from proteins during the roasting process and form complexes with theobromine. In general, when animal and vegetable proteins (among other casein, soja protein, zein, gliadin) are subjected to brief, dry heating to a temperature of 260°C, the produce bitter, aqueous extracts, whose taste threshold values (0.0005 %-0.0008 %) lie within the range of the average values quoted for quinine hydrochloride (0.001 %) (3). The bitter taste which is produced as a result of enzymatic protein hydrolysis (4) must be considered as being negative. This bitter taste may, for instance, occur as an off - flavour during the maturing of cheese, and also prevents a wider

0-8412-0526-4/79/47-115-093$09.75/0

use of enzymatic processes of proteolysis in the food industry. BAUR et al. (5) were recently able to demonstrate that the oxidation of fats can lead to bitter compounds. This means that an important cause of the bitter off-flavour which occurs in many foodstuffs of vegetable origin such as oats or legumes has probably been discovered.

For the reasons mentioned above considerable interest is taken in trying to gain an understanding of sensory qualities through the structure of the compounds involved. Many attempts have been made to derive general rules for relations between chemical structure and taste, but so far it has not been possible in general to explain sensory properties satisfactorily on the basis of structure or to make safe predictions regarding them.

If we wish to find answers to these problems we must take the simplest possible compounds as the basis of our initial investigations and use them to determine which structural elements are connected with specific sensory qualities and what sensory consequences result when these structural elements are altered in specially selected ways.

An explanation of these relations will follow now, taking some of the sweet and bitter compounds as our examples.

Sweet taste is produced by a wide variety of compounds. SHALLENBERGER and ACREE (6) regard an acid/base system (AH/B-System) as the shared structural element. This system must satisfy certain steric conditions and can interact with a complementary system of a receptor via 2 hydrogen bonds (Fig. 1). KIER (7) expanded this model by assuming an additional interaction with an apolar group X in a suitable position (Fig. 1). Both models are applicable to compounds with great variations in structure. There are no similar comprehensive concepts for bitter compounds which can also occur in the most varying chemical classes.

Amino acids and related compounds, as well as simple aromatic compounds, are very well suited for developing further models, as in these chemical classes sweet and bitter taste and also sweet/bitter taste occur. Thus we have the opportunity of dealing with the taste qualities sweet and bitter uniformly.

Detailed investigations of structure and taste require quantitative data on the sensory side. The taste recognition threshold value is especially well suited for this purpose as, according to BEIDLER (8), it is connected with the association constant of the stimulus-receptor-complex.

With amino acids the occurrence of sweet taste (9) depends both on the presence of an ammonium group and on the presence of a carboxylate group, corresponding to a bipolar (electrophilic/nucleophilic) contact with a receptor (Table I). The relative position of the two polar groups is important. Sweet taste decreases strongly on transition from 2-aminocarboxylic acids to 4-aminocarboxylic acids (Table II).

Some of the 2-hydroxycarboxylic acids are also sweet (10). However, with these compounds, sweet taste apparently depends to a

Table I

Sweet taste of amino acids: dependence on ammonium and carboxylate groups (9).

$$H-\underset{R}{\overset{Y}{C}}-X$$

R	X	Y	Sweet taste	c_{tsw}(mmol/l)[a]
CH_3	NH_3^+	COO^-	+	12 - 18
CH_3	H	COO^-	-	
CH_3	OH	COO^-	-	
CH_3	Cl	COO^-	-	
H	NH_3^+	COO^-	+	25 - 35
H	$NH_2CH_3^+$	COO^-	+	15 - 20
H	$N(CH_3)_3^+$	COO^-	+	30 - 50
H	$NH_2C_6H_5^+$	COO^-	-	
H	$NHCOCH_3$	COO^-	-	
$(CH_3)_2CHCH_2$	NH_3^+	COO^-	+	2 - 5
$(CH_3)_2CHCH_2$	NH_3^+	H	-	
H	NH_3^+	CH_2OH	-	
H	NH_3^+	$COOCH_3$	-	
$C_6H_5CH_2$	NH_3^+	COO^-	+	1 - 3
$C_6H_5CH_2$	NH_3^+	$COONH_2$	-	

[a] recognition threshold concentration for sweet taste

Table II

Sweet taste of amino acids: dependence on the relative positions of ammonium and carboxylate groups (9)

Amino acid	Sweet taste	c_{tsw}(mmol/l)
D-2-amino-propionic acid	+	12 - 18
3-amino-propionic acid	+	1000 - 1400
D-2-amino-butyric acid	+	12 - 16
D,L-3-amino-butyric acid	+	100 - 300
4-amino-butyric acid	-	

Table III

Sweet taste of D,L-2-hydroxycarbonic acids, $R\text{-}CR_1(OH)\text{-}COO^-$ (10,11)

R	R_1	Sweet taste	c_{tsw}(mmol/l)
H	H	- (100)[+)]	
CH_3	H	- (100)	
CH_3	CH_3	- (100)	
C_2H_5	H	- (100)	
C_3H_7	H	- (100)	
$(CH_3)_2CH$	H	+	15 - 20
$(CH_3)_2CH\text{-}CH_2$	H	+	3 - 5
C_6H_{13}	H	- (100)	
C_6H_5	H	- (100)	

[+)] Maximum concentration (mmol/l) tested

much greater extent on the side chain than is the case with the amino acids. At all events, in the homologous series examined, sweet taste could only be observed in the hydroxy analogues of valine and leucine (Table III) (11).

Contrary to sweet taste, the occurrence of bitter taste only depends on the ammonium group, corresponding to an electrophilic contact (Table IV). On transition from the amino acid to the corresponding amine, c_{tbi} decreases.

The length of the side chain R is important both for the quality and for the intensity of taste (Table V). Up to R=Et, D- and L-amino acids are sweet. R > Et causes bitter taste of increasing intensity in the L-series and increasing sweet taste in the D-series. N-Acylation or esterification abolishes the sweet taste but increases the bitter taste (Table VI).

Thus it must follow that, in the case of peptides, irrespective of the configuration of the amino acids involved, only bitter taste can be expected if the other preconditions (hydrophobic side chains) are satisfied. The examples investigated confirm this assumption: quality and intensity of taste do not depend on the configuration (Table VII). Intensity also seems to be independent of the sequence (Table VIII).

With peptides, the threshold values depend on the amino acids involved. Tables IX and X demonstrate this on the basis of some of the dipeptides and tripeptides.

Qualitative predictions concerning the degree of bitterness to be expected in peptides are possible with the aid of NEY's Q-value (12) (Table XI), which is based on TANFORD's studies (13). Using the ΔF_t values of peptides, quantitative assessment of the expected range of threshold values is also possible on the basis of the amino acid composition (Table XI) (14).

The sweet dipeptide esters of the L-aspartic acid and the L-amino malonic acid (15-21) are interesting exceptions to the bitter taste shared by all other members of the peptide series. Fig. 2 shows that here the amino groups and the free carboxylate groups of the side chains form an electrophilic/nucleophilic-hydrophobic system corresponding to that of a sweet D-amino acid (22).

Fig. 3 demonstrates, on the basis of the sweet and bitter taste of the amino acids, that not only hydrophobicity, but also the shape of the side chains influences the threshold value.

The shape of the side chain is significant for the quality and intensity of taste, as illustrated by the isomeric leucines (Table XII).

Very bulky side chains seem to prevent taste impressions altogether. The influence of polar substituents in the side chain depends very much on their type and position (Table XIII).

Table XIV (23) compiles the results of the investigations conducted with the amino acids with cyclic side chains (1-amino-cycloalkane-1-carboxylic acids).

C_{tsw} decreases greatly as the ring size increases (II,III,IV) and passes through a minimum at the cyclohexane derivative IV. The

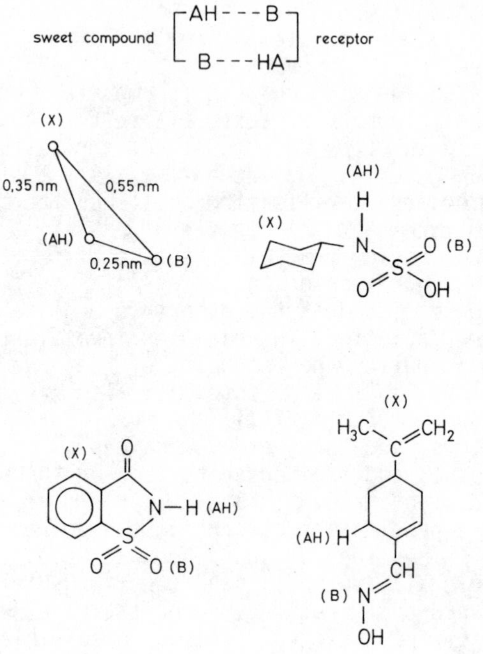

Figure 1. Essential structure elements of sweet compounds according to (6) and (7)

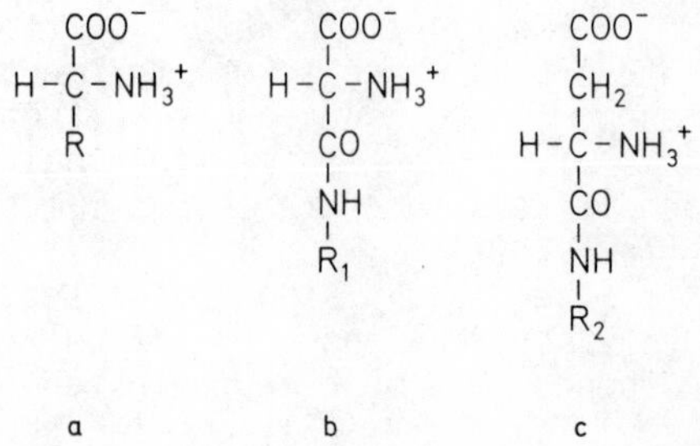

Figure 2. Structures of sweet dipeptide esters: (a) D*-amino acid; (b) peptide of* L*-aminomalonic acid; (c) peptide of* L*-aspartic acid*

Table IV

Bitter taste of amino acids and related compounds: dependence on the ammonium group (24)

$$X-\underset{R}{\overset{Y}{C}}-H$$

R	X	Y	Bitter taste	c_{tbi} (mmol/l)[a]
$CH_3-CH_2-CH_2-CH_2$	NH_3^+	COO^-	+	18 - 22
$(CH_3)_2CH-CH_2$	NH_3^+	COO^-	+	11 - 13
$C_6H_5-CH_2$	NH_3^+	COO^-	+	5 - 7
$4-HO-C_6H_4-CH_2$	NH_3^+	COO^-	+	4 - 6
$C_6H_5-CH_2$	$NH-COCH_3$	COO^-	+	10 - 12
$CH_3-CH_2-CH_2-CH_2$	H	COO^-	-	
$CH_3-CH_2-CH_2-CH_2$	OH	COO^-	-	
$(CH_3)_2CH-CH_2$	NH_3^+	H	+	3 - 4
$4-HO-C_6H_4-CH_2$	NH_3^+	CH_2OH	+	5 - 7
$4-HO-C_6H_4-CH_2$	NH_3^+	$COOC_2H_5$	+	4 - 5
$4-HO-C_6H_4-CH_2$	NH_3^+	$CONH_2$	+	4 - 5

[a] recognition threshold concentration for bitter

Table V

Quality and intensity of taste of amino acids: dependence on the length of the side chain (9, 24)

$$^{+}H_3N - \overset{\displaystyle COO^-}{\underset{\displaystyle R}{C}} - R_1$$

R	R_1	taste quality	c_{tsw} (mmol/l)	c_{tbi} (mmol/l)
H	H	sweet	25 - 35	
CH_3	H	sweet	12 - 18	
H	CH_3	sweet	12 - 18	
CH_3	CH_3	sweet	5 - 10	
C_2H_5	H	sweet/bitter	12 - 16	95 - 100
H	C_2H_5	sweet	12 - 16	
C_3H_7	H	bitter		45 - 50
H	C_3H_7	sweet	3 - 5	
C_6H_5-CH_2	H	bitter		5 - 7
H	C_6H_5-CH_2	sweet	1 - 3	

Table VI

Taste of amino acids: derivatives of ammonium and carboxylic groups (9, 24)

Amino acid/derivative	Taste quality	c_{tsw} (mmol/l)	c_{tbi} (mmol/l)
Gly	sweet	25 - 35	
N-Bz- Gly	bitter		4 - 6
L-Ala	sweet	12 - 18	
N-Bz-L-Ala	bitter		4 - 6
D-Ala	sweet	12 - 18	
N-Bz-D-Ala	bitter		4 - 6
L-Phe	bitter		5 - 7
N-Ac-L-Phe	bitter		10 -12
L-Phe-OMe	bitter		3 - 5
N-Ac-L-Phe-OEt	bitter		1 - 2
D-Phe	sweet	1 - 3	
N-Ac-D-Phe	neutral		
D-Phe-OMe	bitter		3 - 4
D-Phe-NH_2	bitter		2 - 3
N-Ac-D-Phe-OMe	bitter		1 - 2

Ac: acetyl, Bz: benzoyl, Me:methyl, Et: ethyl

Table VII

Taste of peptides: dependence on the configuration of the amino acids (14)

Peptide A-B	A-L-B/L-A-L-B	A-D-B/L-A-D-B	D-A-D-B
Gly-Leu	bitter (19-23)[+)]	bitter (20-23)	
Gly-Phe	bitter (15-17)	bitter (15-17)	
Leu-Leu	bitter (4- 5)	bitter (5- 6)	bitter (5-6)

+) c_{tbi} (mmol/l)

Table VIII

Taste of peptides: dependence on the amino acid sequence (14)

Peptide	c_{tbi} (mmol/l)
Ala-Leu	18 - 22
Leu-Ala	18 - 21
Gly-Leu	19 - 23
Leu-Gly	18 - 21
Ala-Val	60 - 80
Val-Ala	65 - 75
Phe-Gly	16 - 18
Gly-Phe	15 - 17
Phe-Gly-Phe-Gly	1.0- 1.5
Phe-Gly-Gly-Phe	1.0- 1.5

Table IX

Taste of dipeptides A-B: dependence on the amino acid composition (14)

	B	Asp	Glu	Asn	Gln	Ser	Thr	Gly	Ala	Lys	Pro	Val	Leu	Ile	Phe	Tyr	Trp
A										85*	26	21	12	11	6	5	5
Gly											45	75	21	20	16	17	13
Ala												70	20				
Pro	26												6				
Val	21							65	70			20	10				
Leu	12							20	20				4,5			3,5	0,4
Ile	11	43	43	33	33	33	33	21	21	23	4	9	5,5	5,5			0,9
Phe	6							17			2		1,4		0,8	0,8	
Tyr	5												4				
Trp	5		28														0,3

*) c_{tbi} (mmol/l)

Table X

Taste of tripeptides A-B-C: dependence on the amino acid composition (14)

A - B \ C		Gly	Leu	Tyr
A - B			12*)	5
Gly-Gly			75	
Leu-Glu			10	
Ile-Glu	43			
Leu-Gln			3	
Ile-Gln	33			
Gly-Leu	21	55		3
Leu-Gly	20	75	6	
Leu-Val			2	
Val-Leu	10			
Leu-Leu	4.5		1.4	

*) c_{tbi} (mmol/l)

Table XI

Hydrophobicity of amino acids and peptides

TANFORD, NOZAKI (13)

$$\Delta F_t = RT \ln \frac{N_W \quad \gamma W}{N_{Org} \quad \gamma Org}$$

$$H\Phi_i = \Delta F_t \text{ (amino acid}_i) - \Delta F_t \text{ (glycine)}$$

NEY (12)

$$Q \text{ (peptide}_n) = \frac{\Sigma H \Phi_i}{n}$$

WIESER, BELITZ (14)

$$\Delta F_t(\text{peptide}_n) = \Delta F_t(\text{glycine}_n) + \sum_i H\Phi_i$$

Table XII

Taste of amino acids: dependence on the shape of the side chain (9, 23)

Amino acid	R	Taste quality	c_{tsw}(mmol/l)
D-nor-Leu	CH_3-CH_2-CH_2-CH_2	sweet	5 - 8
D-Leu	$(CH_3)_2$CH-CH_2	sweet	2 - 5
D-Ile	CH_3-CH_2-CH(CH_3)	sweet	8 - 12
D,L-tert-Leu	$(CH_3)_3$C	neutral	

Table XIII

Taste of amino acids: dependence on the substituents of the side-chain (9, 24)

Amino acid	Taste quality	c_{tsw} (mmol/l)	c_{tbi} (mmol/l)
D-alanine	sweet	12 - 18	
D-serine	sweet	30 - 40	
L-2-amino-butyric acid	sweet	12 - 16	
L-threonine	sweet	35 - 45	
L-homoserine	sweet	25 - 30	
D-phenylalanine	sweet	1 - 3	
D-tyrosine	sweet	1 - 3	
L-proline	sweet/bitter	25 - 40	25 - 27
L-4-hydroxyproline	sweet	5 - 7	
L-allo-4-hydroxyproline	neutral		

Table XIV

Taste of 1-amino-cycloalkane-1-carboxylic acids (23).

Nr.	Cycloalkane	c_{tsw}(mmol/l)	c_{tbi}(mmol/l)
II	cyclobutane	20 - 30	
III	cyclopentane	3 - 6	95 - 100
IV	cyclohexane	1 - 3	45 - 50
V	cycloheptane	2 - 4	13 - 15
VI	cyclooctane	2 - 4	2 - 5
VII	cyclononane	n.s.(50)$^{+)}$	20 - 50
VIII	cycloundecane	n.s.(20)	n.b.(20)
IX	cyclododecane	n.s.(20)	n.b.(20)
X	2-methylcyclohexane$^{++)}$	n.s.(20)	n.b.(20)
XI	3-methylcyclohexane$^{++)}$	2 - 5	2 - 4
XII	4-methylcyclohexane	8 - 10	1 - 3
XIII	4-ethylcyclohexane	n.s.(50)	1 - 3
XIV	4-tert.-butylcylohexane	n.s.(50)	n.b.(50)
XV	2-amino-norbornane-2-carboxylic acid	ca. 50	5 - 7

$^{+)}$n.s.: not sweet, n.b.: not bitter, in parenthesis maximum concentration (mmol/l) tested

$^{++)}$mixture of isomers

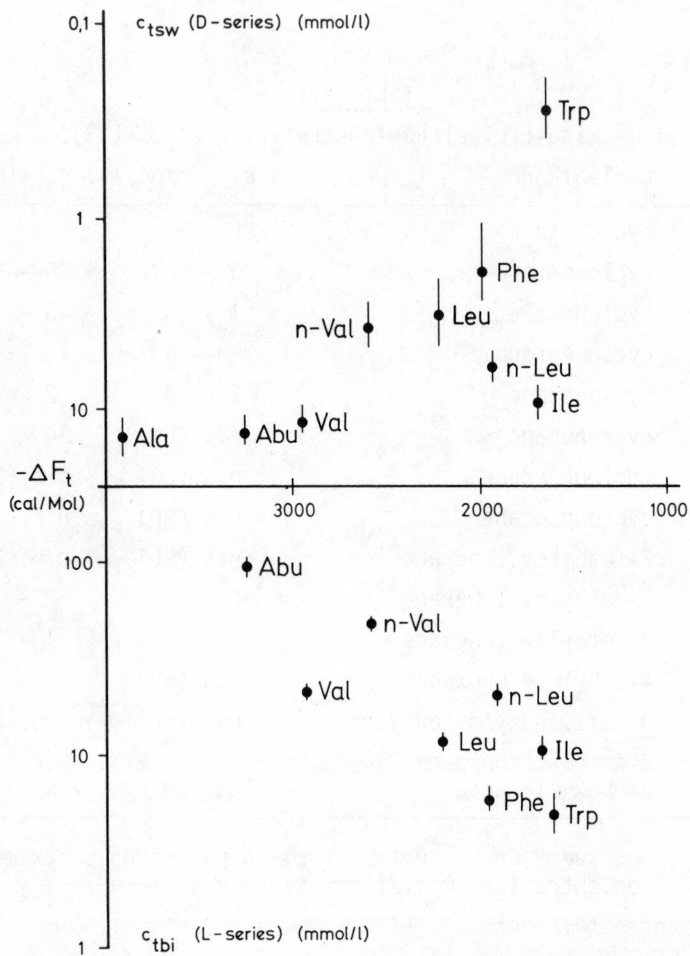

Figure 3. Taste intensity and hydrophobicity of amino acid

following two homologues (V, VI) have the same, somewhat higher threshold value, whereas the remaining members of the series investigated, (VII, VIII, IX), are no longer sweet up to concentrations of 20 resp. 50 mmol/l.

The influence of substituents was tested on the cyclohexane derivative IV. In position 2 (X) a methyl group abolishes the sweet taste, whereas it is tolerated in positions 3 (XI) and 4 (XII), with an increase in c_{tsw} however. Larger substituents,such as the ethyl group (XIII) and the tert. butyl group (XIV) abolish the sweet taste even in position 4.

In the case of the 1-amino-cycloalkane-1-carboxylic acids, the threshold values for bitter taste also pass through a minimum, which, however, in contrast to c_{tsw}, shifts from the six-membered ring to the eight-membered ring. It is also significant that there is a relatively large decrease in c_{tbi} on transition from V to VI and then an immediate increase with VII, whereas, within the range of the compounds III-VI, c_{tsw} goes through a relatively flat minimum.

With the alkyl-substituted cyclohexane derivatives there are also variations in the course of c_{tsw} and c_{tbi}. The 2-methyl compound (X) is not bitter up to a concentration of 20 mmol/l, whereas a 3-methyl group (XI) or a 4-methyl group (XII) decreases c_{tbi} in comparison with the unsubstituted compound IV. Whereas a 4-ethyl group (XIII) abolishes sweet taste, it lowers c_{tbi} even further, in comparison with the methyl compound XII. The 4-tert.-butyl derivative XIV is neither sweet nor bitter.

The bicyclic 2-aminonorbornan-carboxylic acid (XV) is sweet and bitter. Significant, too, is the great difference between c_{tsw} and c_{tbi} which is almost one decimal power. Altogether it can be seen that the steric conditions for bitter compounds are not as limiting as those for sweet compounds.

The investigations mentioned above of amino acids R-$CHNH_2$-COOH (9,24) have shown that D-amino acids are sweet, with c_{tsw} decreasing as the length of R increases. In the L-series the compounds are only sweet up to R $\leq$ Et; from R $\geq$ Et onwards they are bitter, with c_{tbi} decreasing as the length of R increases. As can be expected from these results sweet and bitter taste occur from the 5-membered ring onwards in the 1-amino-cycloalkane-1-carboxylic acids tested here. The 4-methyl and 4-ethyl-cyclohexane derivatives XII and XIII show that sweet taste only occurs up to a side chain lenght of approx. 0.6 nm (as measured from the centre of the C-atom 1 of the ring in the direction of the main axis). This condition would also be satisfied in the case of the cyclooctane derivative (VI) and the cyclononane derivative (VII) but the latter is nevertheless not sweet. Apparently the width of the ring in the region of the C-atoms 5 and 6 (cf. Fig. 4) prevents contact with the receptor in the case of VII. From cyclodecane derivative onwards the side chain length of 0.6 nm is also exceeded. The results would suggest a barrier at the receptor. PAUTET and NOFRE (25) were able to demonstrate with N-alkyl-sulfamates

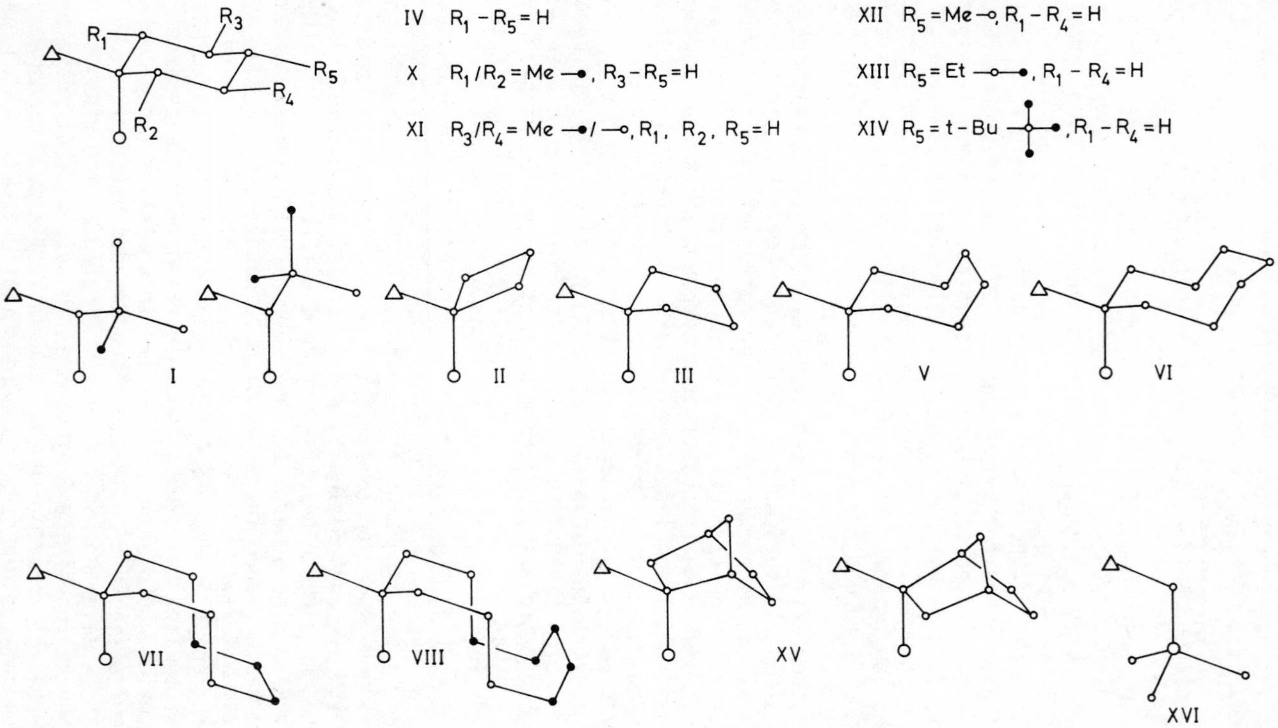

IV $R_1 - R_5 = H$

X $R_1/R_2 = Me$ —●, $R_3 - R_5 = H$

XI $R_3/R_4 = Me$ —●/—○, R_1, R_2, $R_5 = H$

XII $R_5 = Me$ —○, $R_1 - R_4 = H$

XIII $R_5 = Et$ —○—●, $R_1 - R_4 = H$

XIV $R_5 = t\text{-}Bu$, $R_1 - R_4 = H$

Figure 4. Structures of some amino acids (selected probable isomers/conformers): (○) positions allowed and (●) forbidden for atoms HH in sweet compounds are marked; (○) NH_3^+; (△) COO^-; roman numbers as in Table XIV except (I) leucine; and (XVI) N-*trimethylglycine* (*cf.* (30–35))

that sweet taste does not occur in these compounds if the length of 0.7 nm, as measured from the centre of the N-atom, is exceeded. For the receptors of alkyl sulfamates and amino acids it can therefore be assumed that there is a barrier at approximately the same distance from the bonding site for nitrogen.

The 2-methylcyclohexane derivative X, the norbornane derivative XV, the tert.leucine (I) and the N-trimethylglycine (XVI) provide information about the structural preconditions for sweet taste in the vicinity of ammonium and carboxylate groups (cf. Fig. 4 and 5).

D-Valine is sweet (9) but I and X are not. The C-atom 3 of a sweet amino acid may therefore only carry a maximum of two methyl groups. If we assume that, in the case of X, the isomers represented in Fig. 4 with the ammonium group in axial position and with the carboxylic- and methyl-groups in equatorial position predominate, then it is also clear which of the three possible positions for a methyl group at the C-atom 3 is incompatible with the occurrence of sweet taste. The sweet taste of XV suggests that a methyl group in axial position would have to be allowed (cf. Fig. 5). The sweet taste of N-trimethylglycine (XVI) demonstrates, that an acidic hydrogen is not essential, but that an electrophilic group as $-N(CH_3)_3^+$ is sufficient.

In Fig. 5 the structural formulae of some compounds that were investigated have been superimposed, in order to throw into relief those position whose occupation is allowed for sweet taste and those whose occupation is forbidden for sweet taste. The bonding site of the receptor stands out clearly in relief as a hydrophobic pocket with 2 polar contact points.

The ring size has a determining effect on the threshold values not only in the case of the sweet-bitter 1-amino-cycloalkane-1-carboxylic acids but also, as our investigations have shown, in the case of other cyclic compounds, as for example the bitter cycloalkanones, azacycloalkanes, lactams and lactones (Table XV), as well as the sweet cycloalkanesulfamates (Table XVI). From the tables it follows that the hydrophobic contact with the receptor apparently attains a maximum in general with ring sizes 6-8 on the part of sweet as well as bitter compounds.

With simple aromatic compounds (Tables XVII and XVIII) the effect of structure on sensory qualities is easily recognizable, too. In order to determine electrophilic and nucleophilic centers in aromatic compounds their charge distributions were calculated. The values for the atomic charges were obtained from a computer program which is based on a model for partial equalization of orbital electronegativity (26, 27). This model was extended to π-systems (28).

Phenol is sweet and bitter (XVII) as was definitely established despite a strong by-taste. From the charge distribution follows that the hydrogen of the HO-group and the carbon in 2-position, or the hydrogen in 2-position and the oxygen can function as an electrophilic/nucleophilic system (Fig. 17). Additio-

Table XV

Taste of selected cyclic compounds (11)

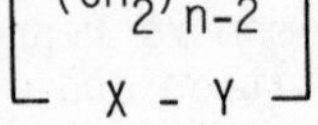

Structure: ring of $(CH_2)_{n-2}$ closed by X - Y

X - Y	n	c_{tbi} (mmol/l)
CH_2-CO	5	10 - 15
	6	6 - 10
	7	5 - 6
	8	2 - 4
	12	n.b. (3)
CH_2-NH	6	8 - 10
	7	0.6- 1
	8	0.5- 1.2
NH - CO	5	6 - 8
	6	3 - 4
	7	4 - 8
	8	12 - 15
	9	10 - 20
O - CO	5	50 - 60
	6	10 - 20

Table XVI

Taste of cycloalkanesulfamates (11)

Structure: ring of $(CH_2)_{n-1}$ closed by CH - NH - SO_3H

n	c_{tsw} (mmol/l)
4	100
5	2 - 4
6	1 - 3
7	0.5 - 0.7
8	0.5 - 0.8

Table XVII

Taste of hydroxy- and amino-benzenes (11)

Nr.	R_2	R_3	R_4	R_5	c_{tsw} (mmol/l)	c_{tbi} (mmol/l)
XVII	H	H	H	H	5 - 15	40 - 60
XVIII	HO	H	H	H	n.s. (50)+)	5 - 10
XIX	H	HO	H	H	5 - 7	5 - 15
XX	H	H	HO	H	15 - 20	15 - 25
XXI	HO	HO	H	H	n.s. (60)	6 - 8
XXII	HO	H	HO	H	n.s. (100)	-
XXIII	H	HO	H	HO	0.5 - 1	5 - 15
XXIV	HO	H	Me	H	n.s. (100)	
XXV	Me	HO	H	H	n.s. (100)	5 - 15
XXVI	H	HO	Me	H	+	5 - 15
XXVII	H	HO	H	Me	3 - 7	10 - 15
XVIII	H	H	H	H	n.s. (50)	+
XXIX	NH_2	H	H	H	n.s. (100)	40 - 60
XXX	H	NH_2	H	H	n.s. (100)	n.b.

+) n.s.: not sweet, n.b.: not bitter, maximum concentration (mmol/l) tested in parenthesis

Table XVIII

Taste of benzoic acids and of nitrobenzenes (11)

Nr.	R_2	R_3	R_4	R_5	R_6	c_{tsw} (mmol/l)	c_{tbi} (mmol/l)
XXXI	H	H	H	H	H	10 - 15	n.b.+)
XXXII	COO^-	H	H	H	H	n.s. (50)	
XXXIII	H	H	COO^-	H	H	n.s. (40)	
XXXIV	H	COO^-	H	COO^-	H	n.s.(100)	
XXXV	HO	H	H	H	H	3 - 5	n.b.
XXXVI	H	HO	H	H	H	5 - 7	
XXXVII	H	H	HO	H	H	n.s.(100)	8 - 12
XXXVIII	HO	H	HO	H	H	1 - 2	+ ?
XXXIX	HO	H	H	H	HO	n.s. (50)	
XL	H	HO	H	HO	H	n.s.(100)	+ ?
XLI	NH_2	H	H	H	H	3 - 5	n.b.
XLII	H	NH_2	H	H	H	8 - 10	n.b.
XLIII	H	H	NH_2	H	H	n.s.(100)	30 - 40
XLIV	H	NH_2	H	NH_2	H	- (50)	
XLV	Cl	H	H	H	H	n.s. (50)	
XLVI	H	Cl	H	H	H	-	n.b. (50)
XLVII	H	H	Cl	H	H	-	4 - 6
XLVIII	Cl	H	Cl	H	H	n.s.(100)	+
IL	Cl	H	H	H	Cl	n.s. (50)	+
L	NO_2	H	H	H	H	0.5 - 1	2 - 6
LI	H	NO_2	H	H	H	5 - 8	0.8- 1
LII	H	H	NO_2	H	H	7 - 9	0.1- 0.3
LIII	H	H	H	H	H	0.3 - 0.5	
LIV	HO	H	H	H	H	2 - 6	
LV	H	HO	H	H	H	0.5 - 1	
LVI	H	H	HO	H	H	5 -15	

COO^- (benzene ring with substituent R)

NO_2 (benzene ring with substituent R)

+) cf. table XVII

nal hydroxyl groups in positions 3 (XIX) and 5 (XXIII) intensify the negative charge in 2-position, but weaken it on the oxygen (Fig. 17) and result in a decrease in c_{tsw}. This leads to the conclusion that of the 2 possibilities mentioned for the electrophilic/nucleophilic system the first one will probably prevail. The effect of additional hydroxyl groups on c_{tbi} is not so great. The hydrogen of the HO-group must be seen as the electrophilic group, its charge in the case of XVII, XIX and XXIII being constant. With hydroquinone (XX), c_{tsw} is higher than with XVII and XIX. As the charge in the 2-position is only very slightly diminished (Fig.17), we can assume disturbances of the hydrophobic contact by an HO-group in 4-position.

Neighbouring hydroxyl groups abolish sweet taste (XVIII, XXI, XXII, XXIV) but are compatible with the bitter taste (XVIII, XXI). A methyl group in 2-position (XXV) also abolishes sweet taste.

Aniline XXVIII and the diamine benzenes XXIX and XXX are not sweet. Apparently the positive charge of the hydrogen atoms of the amino group, which is lower than that of the phenols (Fig.17), is not sufficient for contact with the receptor.

With benzoic acid (XXXI) the hydrogen in 2-position and the oxygen atoms of the carboxyl group must be regarded as the electrophilic/nucleophilic system on the grounds of the charge distribution (Fig.17). Comparison with the phenols in Table XVII shows that the benzene nucleus can contribute the electrophilic as well as the nucleophilic group to the bipolar system. The effect of further substituents depends on type and relative position.

2-Hydroxy-(XXXV) and 2-aminobenzoic acid (XLI) are significantly sweeter than benzoic acid (XXXI). Apparently the bipolar system of these compounds consists of a hydrogen from the HO- resp. NH_2-group and an oxygen from the carboxylate group. In the case of the corresponding 3-substituted benzoid acids (XXXVI,XLII) the same bipolar system can not work for steric reasons. On the other hand the charge in the 2-position of these compounds is not significantly altered in comparison to benzoic acid and the hydrophobicity has decreased. Therefore it is difficult to interprete the relative low values for c_{tsw}.

It seems that hydroxy- and amino-substituents in 4-position to the carboxylate group are not compatible with sweet taste (XXXVII, XLIII). p-Nitrobenzoic acid (LII) however, is sweet. Because of the charge distribution of this compound, bipolar contact with the receptor probably takes place, as with nitrobenzene (LIII) via an oxygen of the nitro group and the hydrogen in 2-position (Fig.17). The hydrophobic contact can be disturbed to varying degrees, depending on position, by the carboxylate group. Therefore the c_{tsw} of all the nitrobenzoic acids tested (L,LI,LII) is higher than the c_{tsw} of nitrobenzene itself (LIII). The same is true for the nitrophenols (LIV-LVI).

With the nitrobenzoic acids L-LII the inverse behaviour of c_{tsw} and c_{tbi} is worthy of note: from the 2-to the 4-derivative c_{tsw} increases whereas c_{tbi} decreases. The amino benzoic acids

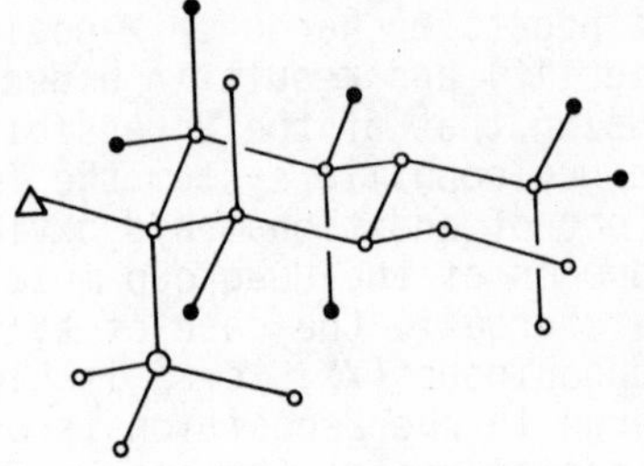

Figure 5. Superposition of some amino acids ((○) allowed and (●) forbidden positions for sweet taste are marked; cf. legend to Figure 4)

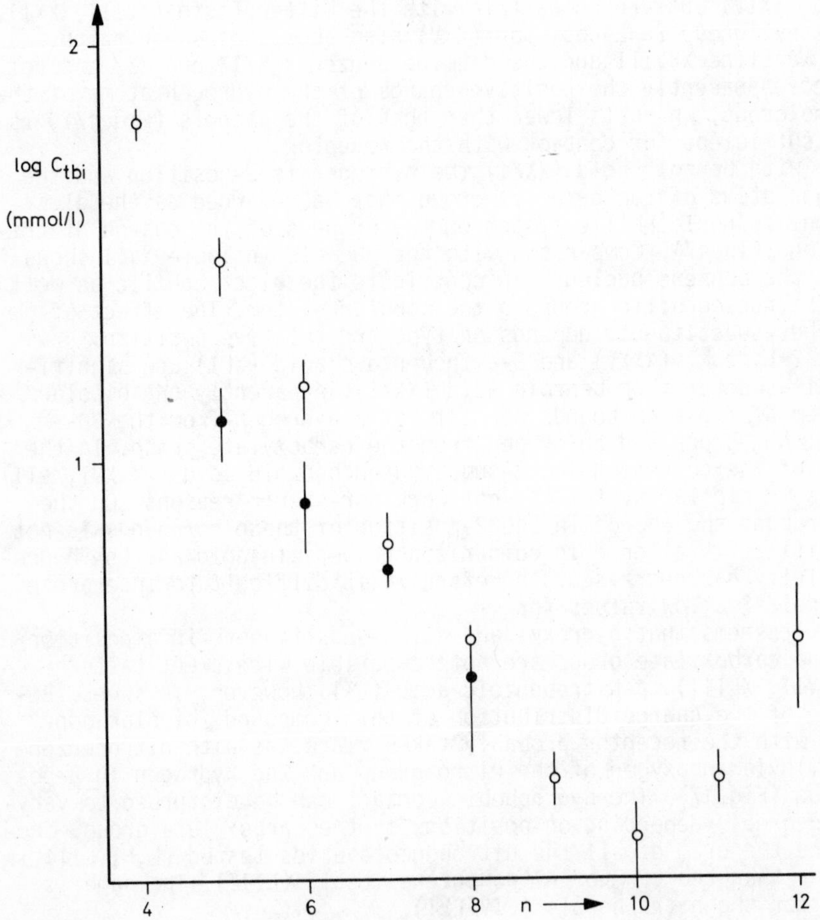

Figure 6. Bitter taste of some ketones: (○) 2-alkanones; (●) cycloalkanones; (n) total number of C-atoms

(XLI-XLIII) and the hydroxy-benzoic acids (XXXV-XXXVII) probably have the same tendency, but, up to the concentrations tested, the 4-derivatives are not sweet and the 2- and 3-derivatives are not bitter.

The predominating view in the literature is that both bitter and sweet compounds need a bipolar system. When investigating the taste of amino acids and related compounds, we came to the hypothesis that for bitter taste a monopolar (electrophilic) hydrophobic structure is sufficient and that in the case of bipolar hydrophobic compounds the overall steric properties determine whether they are sweet, bitter, sweet-bitter or tasteless. For checking our hypothesis we have determined the taste quality and the recognition threshold values of some further simple compounds. Ketones with sufficiently long side chains are bitter (Fig.6). The carbonyl-C-atom is assumed to be the polar (electrophilic) contact group. Open chain and cyclic ketones with 7 or more C-atoms have approximately the same c_{tbi}. Therefore their conformation at the receptor site may be similar. In the case of 2-alkanones c_{tbi}-values seem to pass through a minimum. As mentioned above the values of Table XV show that c_{tbi} of cycloalkanones, oxacycloalkanes azacycloalkanes, lactams and lactones is related to the ring size. In the case of five- and six-membered rings, lactones have significantly higher, lactams significantly lower c_{tbi}-values in comparison with cycloalkanones of the same ring size, while those of azacycloalkanes are nearly equal. On the basis of the charge distribution (Fig.17) the carbonyl-C-atoms in the case of cycloalkanones and lactones, and a hydrogen-atom of the nitrogen in the case of the azacycloalkanes can be assumed to be electrophilic contact groups. With the lactones the ring oxygen probably leads to a disturbance of the hydrophobic contact and thus to an increase in c_{tbi}. With the lactams the carbonyl-C-atom or the hydrogen of the NH-group is the possible polar contact group (Fig.17).

Open chain esters, amides and alkyl-ureas may also be bitter (Table XIX). The c_{tbi}-values depend on the hydrophobicity of the side chains. Primary, secondary and tertiary amines are bitter (Table XX). The c_{tbi}-values of compounds with equal side chains decrease in the same order. But with the i-butyl-residue, c_{tbi} reaches a minimum with the secondary amine. Possibly there are problems with three somewhat bulky residues at the receptor site. c_{tbi} increases with hydrophilic substituents in the side chain, as is shown with the ethyl and hydroxy-ethyl compounds.

The examples of substituted piperidines and pyridines show that c_{tbi} depends on position and polarity of the substituents (Table XXI). Apolar groups seem to make the best fit with the receptor at position 3 or 4, while the negative influence of polar groups on the hydrophobic contact seems to be minimal at position 2.

The homologous compounds tested follow a linear relationship between log c_{tbi} and the number of C-atoms in the side chain R (Fig.7), according to the equation (29):

Table XIX

Taste of amides, esters and alkyl ureas (11)

	R_1	R_2	c_{tbi} (mmol/l)
R_1COOR_2	Phenyl	Ethyl	2 - 5
	4-Hydroxy-phenyl	Methyl	8 - 10
		Ethyl	4 - 6
		Propyl	1.5 - 2.5
	3,4,5- Trihydroxy-phenyl	Propyl	0.8 - 1.0
		Octyl	0.15 - 0.2
$R_1HN-C(=O)-NHR_2$	H	H	60 - 70
	Methyl	H	35 - 40
	Ethyl	H	20 - 25
	Propyl	H	10 - 15
	Butyl	H	5 - 7.5
	Methyl	Methyl	25 - 30
	Ethyl	Ethyl	12.5 - 15
$RCONH_2$	Methyl		n.b.*)
	Ethyl		50 - 55
	Propyl		17.5 - 22.5
	Butyl		17.5 - 22.5
	Phenyl		0.8 - 1

*) n.b.: not bitter

Table XX

Taste of amines (11)

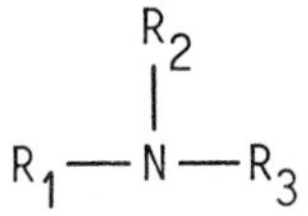

R_1	R_2	R_3	c_{tbi} (mmol/l)
Propyl	H	H	15 - 25
Butyl	H	H	4 - 8
i-Butyl	H	H	4 - 5
Pentyl	H	H	1.5 - 3
Ethyl	Ethyl	H	5 - 15
HO-Ethyl	HO-Ethyl	H	20 - 40
i-Butyl	i-Butyl	H	0.4 - 0.6
Ethyl	Ethyl	Ethyl	2 - 3
HO-Ethyl	HO-Ethyl	HO-Ethyl	10 - 30
i-Butyl	i-Butyl	i-Butyl	0.8 - 2
Benzyl	H	H	2 - 3
Benzyl	Methyl	Methyl	0.6 - 0.9
cyclo-Propyl	H	H	n.b.+)
-Butyl	H	H	35 - 45
-Pentyl	H	H	8 - 10
-Hexyl	H	H	1.5 - 2
-Octyl	H	H	0.5 - 0.7

*) n.b.: not bitter

Table XXI

Taste of piperidines and pyridines (11)

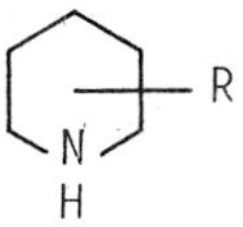

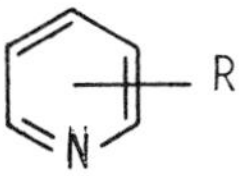

R	c_{tbi} (mmol/l)
H	8 - 10
1-Methyl	10 - 15
2-Methyl	5 - 8
3-Methyl	1 - 4
4-Methyl	1 - 2
4-Phenyl	0.3 - 0.5
2-COOH	25 - 30
2-COOH	5 - 7
3-COOH	20 - 25
$3\text{-}CONH_2$	6 - 8

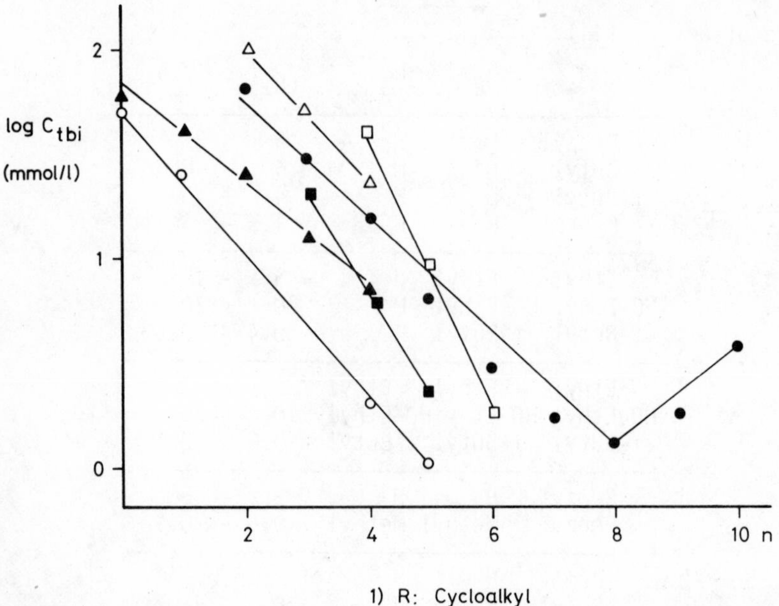

Figure 7. Homologous bitter compounds: relationship between taste intensity and the number of C-atoms in the side chain (●) R-CO-CH_3; (▲) R-NH-CO-NH_2; (△) L-R-CH(NH_2)-COOH; (□) R'-NH_2; (■) R-NH_2; (○) [lactone structure: R–(γ-butyrolactone)=O]

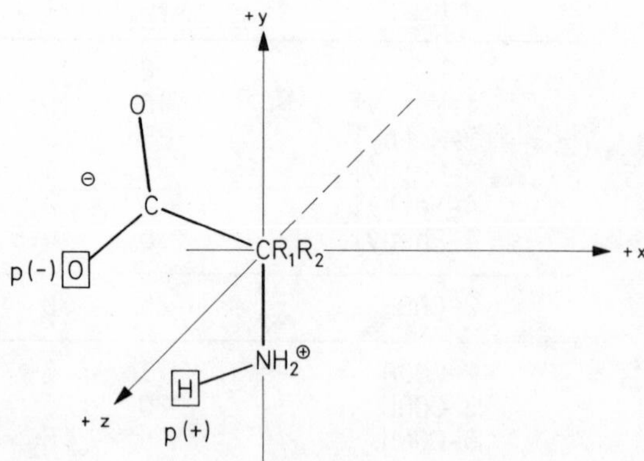

Figure 8. Fixation of sweet and bitter compounds in rectangular coordinates—amino acids

$$\log c_{tbi} = a \cdot n + b$$

As with the 2-alkanones in some cases this seems to be true only up to a limited value of n.

All together the examples show that monopolar (electrophilic) hydrophobic compounds can cause bitter taste and that the intensity of bitter taste depends on the size and shape of the hydrophobic part of the molecule.

Expanding SHALLENBERGER and ACREE's concept for sweet compounds further, we regard not only AH/B-systems capable of forming hydrogen bonds as a potential polar contact grouping but all electrophilic/nucleophilic systems. As a further development of KIER's concept we assume the existence of an expanded hydrophobic contact, not a restricted one.

Size and shape of the hydrophobic part are important for the quality and for the intensity of taste. The individual results can be classified formally in a model. Essential structural elements are

- for sweet compounds 2 polar groups, an electrophilic one p (+), a nucleophilic one p (-). An apolar group "a" is not essential but important for the intensity of sweet taste.
- for bitter compounds one polar group, an electrophilic one p (+) and an apolar group "a".

The coordinate system was so chosen that in the case of 2-amino-carboxylic acids the C-atom 2 is at the origin and the polar groups p (+) and p (-) occupy the positions resulting from Fig.8.

In this system, by superimposing probable conformers of 2-amino-carboxylic acids - as in Fig. 5 - one can indicate positions allowed and forbidden for sweet compounds (Fig.9). Even if there is only information regarding one forbidden position on the +z side and the dotted line is therefore hypothetical, it is still clearly recognizable that the forbidden positions as a whole are not arranged symmetrically to the x-axis.

A D-amino acid, e.g. D-norleucine (sweet), could occupy the positions p (+) p (-) a (+x, ±y, +z) (Fig.10), whereby the expression in brackets after "a" indicates the expansion directions of the apolar group.

A short-chain L-amino acid, e.g. L-alanine (sweet) can occupy the positions p (+) p (-) a (+x, ±y, -z), whereas L-norleucine (bitter) in stretched conformation cannot do so because of forbidden positions (Fig.10), altough it can occupy p (+) a (+x, ±y, +z) after 60° rotation about the y-axis (Fig.10). This rotation about the y-axis is only one possibility, for with monopolar occupation, where fixation of the origin is no longer meaningful, rotation about other axes is also possible, so that this situation must be generally denoted as p (+) a (+x, ±y, ±z).

1-Aminocyclohexane-1-carboxylic acid (sweet/bitter) would occupy the position p (+) p (-) a (+x, ±y, ±z) (Fig.11).

These examples show, that compounds can be arranged with re-

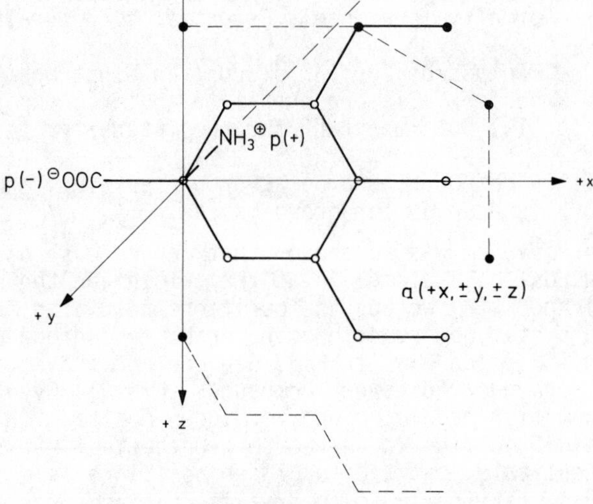

Figure 9. Fixation of sweet and bitter compounds in rectangular coordinates—amino acids: (○) allowed and (●) forbidden positions for sweet taste

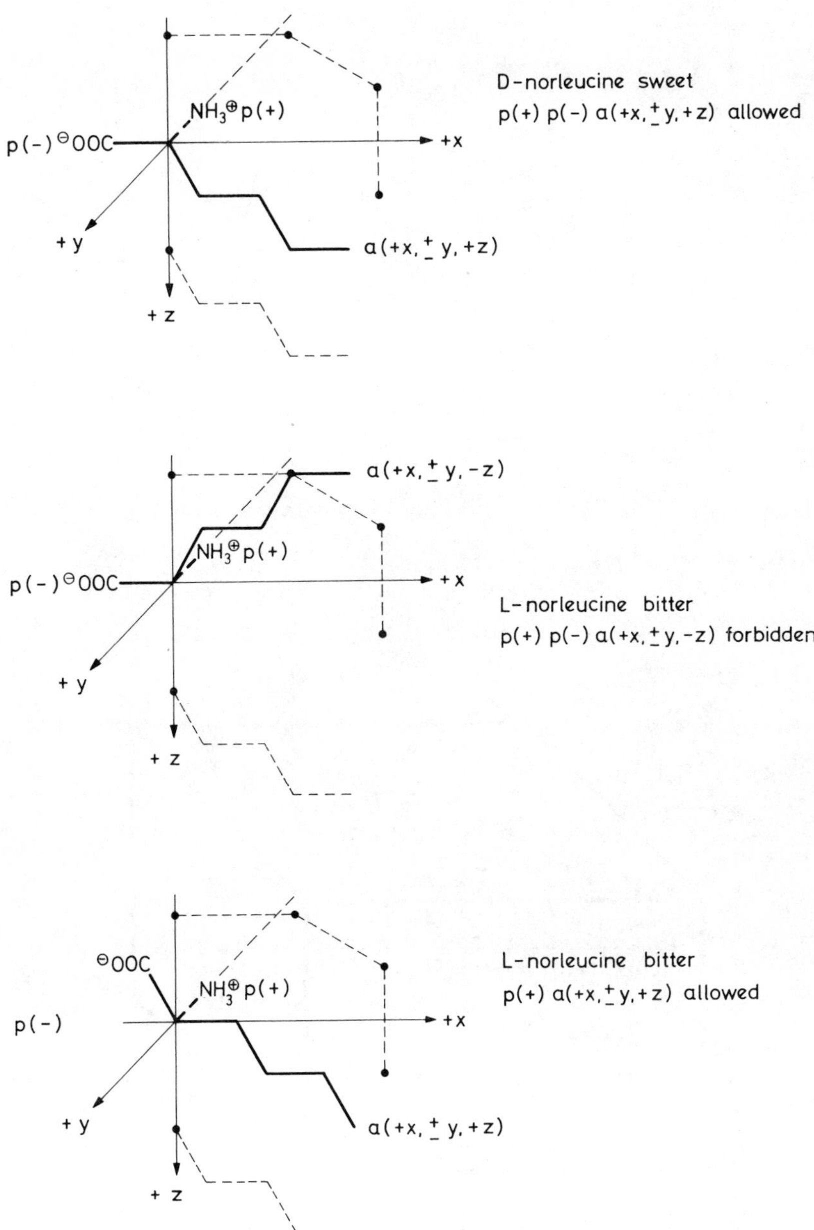

Figure 10. Fixation of sweet and bitter compounds in rectangular coordinates—D-norleucine and L-norleucine in allowed and forbidden positions

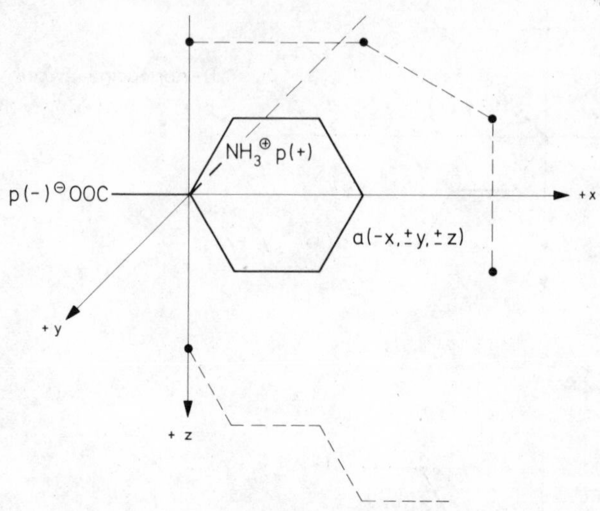

Figure 11. Fixation of sweet and bitter compounds in rectangular coordinates—1-amino-cyclohexane-1-carboxylic acid (sweet/bitter) p(+)p(−)a(+x, ±y, ±z)

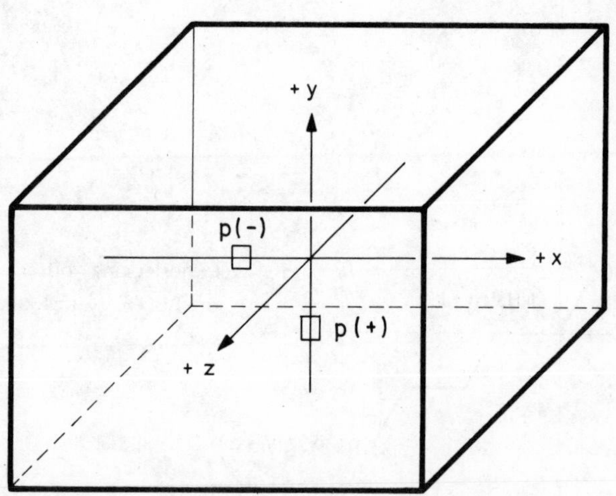

Figure 12. Fixation of sweet and bitter compounds in rectangular coordinates—polar (p(+), p(−)) and apolar (a(±x, ±y, ±z)) occupation possibilities for compounds with the taste qualities sweet: p(+)p(−)a(+x, ±y, + z), p(+)p(−)-a(+x, ±y, −z); sweet–bitter: p(+)p(−)a(+x, ±y, ±z); bitter: p(+)a(±x, ±y, ±z)

gard to their sweet and/or bitter taste qualities according to their varying occupation possibilities(-it is generally assumed that a compound occupies all positions to which it has access -) in the following (Fig.12):

sweet: p(+) p(-) a(+x, $\pm$y, +z) or p(+) p(-) a(+x, $\pm$y, -z)
(In the latter case the C-atom at the coordinate origin may only carry substituents R $\leq$ Et.)
sweet-bitter: p(+) p(-) a(+x, $\pm$y, $\pm$z)
bitter: p(+) a($\pm$x, $\pm$y, $\pm$z)

The aromatic compounds investigated also fit in well with this formal model. The benzene ring is located in the x/y plane, but it extends towards the sides +z and -z with its π-orbitals, so that in the presence of a suitable bipolar system sweet-bitter taste can be expected. The deviations observed, i.e. the occurrence of either sweet or bitter taste on their own (Tables XVII, XVIII) must be traced back to substituents, which either disturb the bipolar system or are located in sterically forbidden positions. The benzoic acids can occupy the positions p(+) p(-) a(+x, $\pm$y) (Fig.13), likewise the nitrobenzenes (Fig.14). For the phenols the corresponding positions would be p(+) p(-) a(-x, $\pm$y) (Fig.15). The other possible arrangement (dotted line in Fig.15) would bring the phenyl residue into a position which in the case of many other compounds (carboxylic acids, nitro compounds) is not occupied by a hydrophobic group but by a polar one. Accordingly a hydrophobic contact in the -x direction too, is not out of the question. This possibility is also advantageous for the placing of bitter peptides and 2.5-dioxopiperazines, which could bring several apolar groups into contact by occupying p(+) a($\pm$x, $\pm$y, $\pm$z).

The general model developed for sweet and bitter compounds leads to a sweet-bitter receptor, which can be given formal representation as a hydrophobic pocket with a bipolar system (Fig. 16). If we designate the contact of a stimulus with one, resp. both polar groups, as monopolar, resp. bipolar, and the hydrophobic contact with one, resp. both sides (+z or -z, resp. $\pm$z) of the pocket, as monohydrophobic, resp. bihydrophobic, then the taste qualities sweet and bitter can be formally traced back to the following contacts:

sweet → bipolar-monohydrophobic
sweet-bitter → bipolar-bihydrophobic
bitter → monopolar-monohydrophobic or bihydrophobic.

This formal system cannot claim to make any statement regarding a real receptor, but it has the advantage of making it possible for the first time, to treat the taste qualities sweet and bitter uniformly and to classify them in one model.

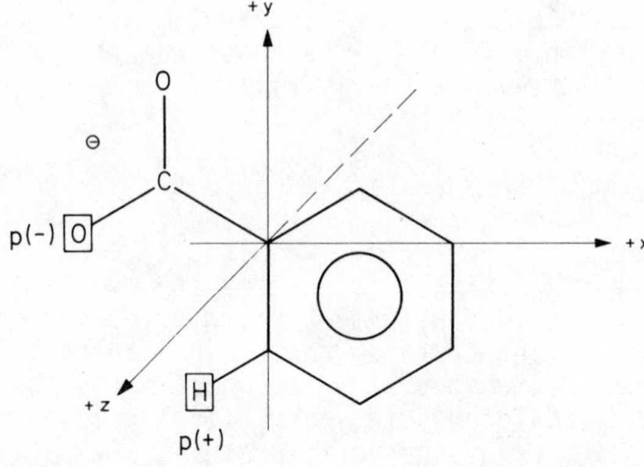

Figure 13. Fixation of sweet and bitter compounds in rectangular coordinates—benzoic acid

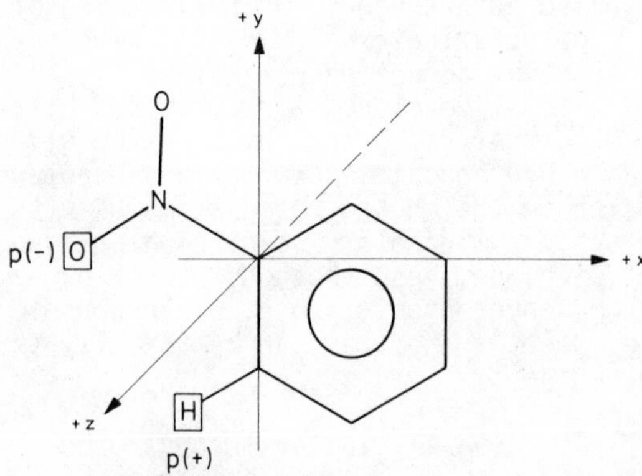

Figure 14. Fixation of sweet and bitter compounds in rectangular coordinates—nitrobenzene

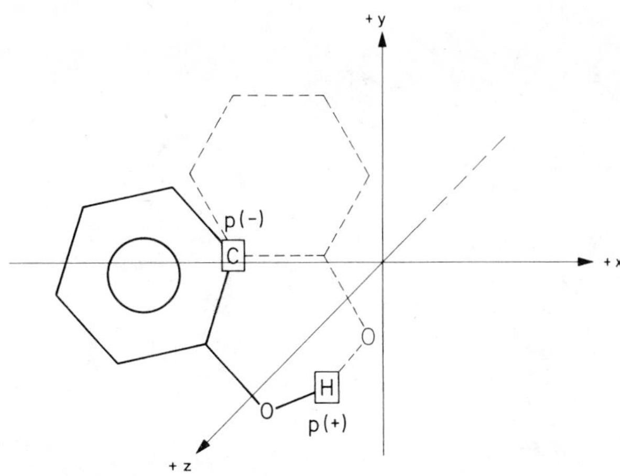

Figure 15. Fixation of sweet and bitter compounds in rectangular coordinates—phenol

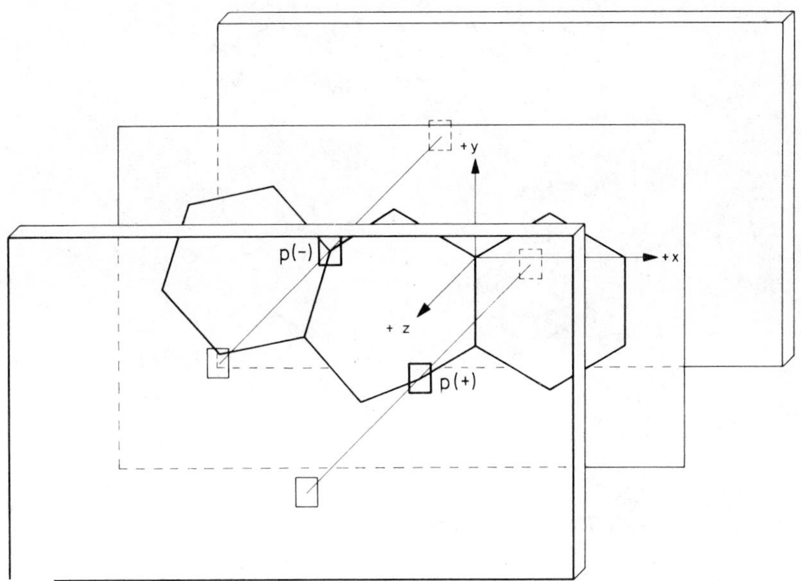

Figure 16. Representation of the hydrophobic pocket and the polar contact groups of a formal sweet–bitter receptor

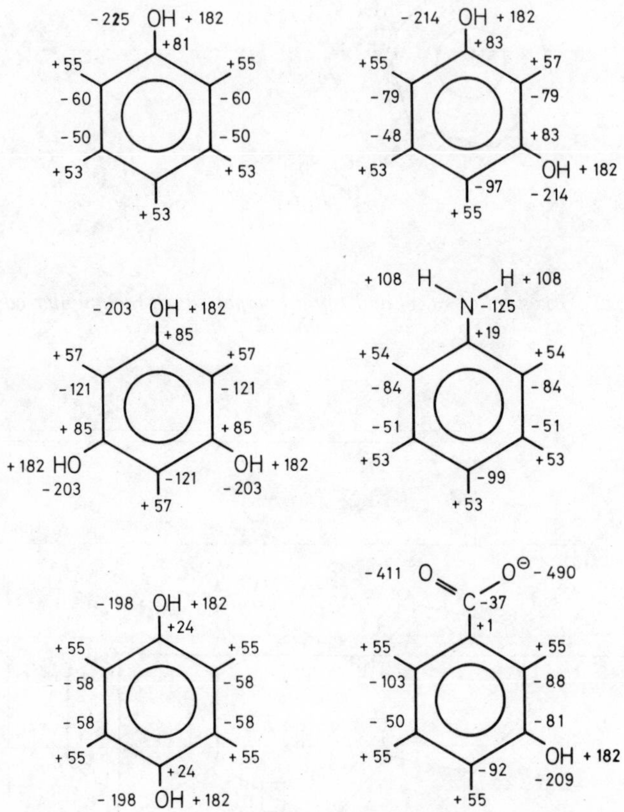

Figure 17. Atomic charge distribution of selected compounds ($e \cdot 10^3$) (26, 27, 28)

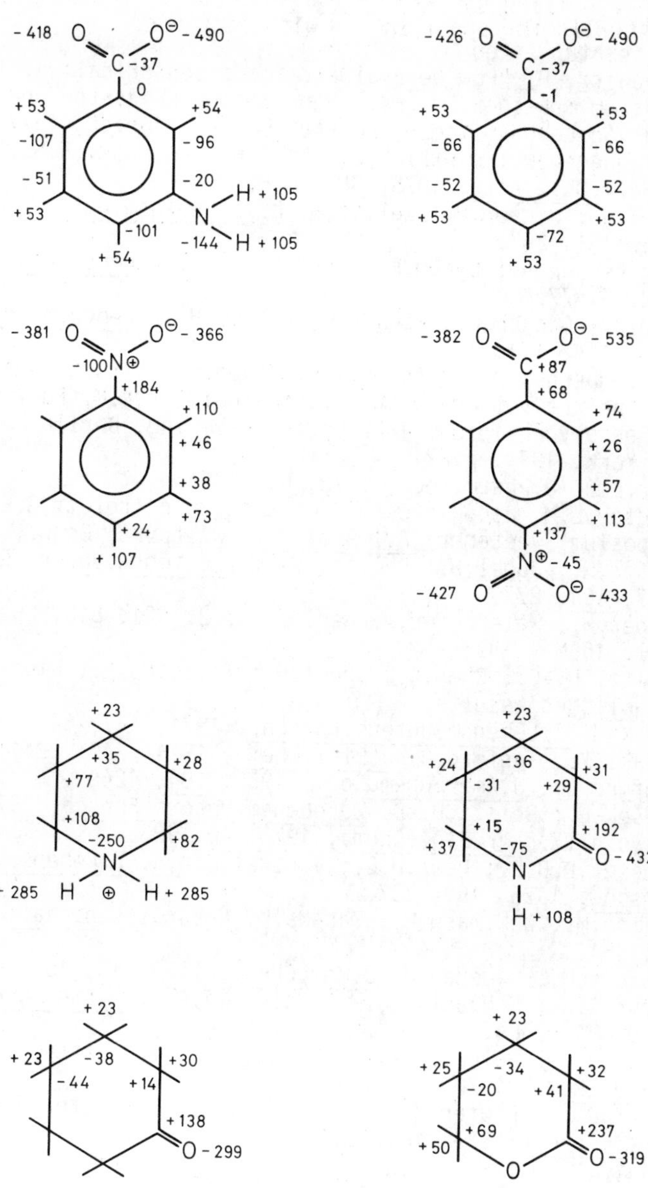

Figure 17. Continued

Acknowledgments. We are indebted to Prof. Dr. Unterhalt, Marburg for supplying us with cycloalkane sulfamates and to Prof. Dr. Tressl, Berlin for supplying us with lactones.

Literatur cited

1. Moskowitz, H. "How we evaluate foods sensorically". In: Raunhardt, O.; Escher,F. (Ed.) Sensorische Erfassung und Beurteilung von Lebensmitteln, Forster Verlag, Zürich, 1977, p.32.
2. Pickenhagen,W.; Dietrich,P.; Keil,B.; Polonsky,F.N.; Lederer, E. Helv.Chim.Acta, 1975, 58, 1078.
3. Jugel,H.; Wieser,H.; Belitz,H.-D. Z.Lebensm.Unters.Forsch., 1976, 161, 267.
4. Petritschek,A.; Lynen,Fr.; Belitz,H.-D. Lebensm.Wiss.Techn., 1972, 5, 77.
5. Baur,C.; Grosch,W.; Wieser,H.; Jugel,H. Z.Lebensm.Unters.Forsch. 1977, 164, 171.
6. Shallenberger,R.S.; Acree,T.E.; Chemical structure of compounds and their sweet and bitter taste. Beidler,L.M.(Ed.) Handbook of Sensory Phyiology 4/2. Springer-Verlag, Berlin-Heidelberg-New York, 1971, p.221.
7. Kier,L.B. J.Pharm.Soc., 1972, 61, 1394.
8. Beidler,L.M. Biophysics of Sweetness. In:Inglett,G.E.(Ed.) Symposium Sweeteners AVI Publ.Co.: Westport, Conn., 1974, p.10.
9. Wieser,H.; Jugel,H.; Belitz,H.-D. Z.Lebensm.Unters.Forsch. 1977, 164, 277.
10. Wagner,H.; Maierhofer,A. Ger.Offen. 2, 521, 816, 25 Nov. 1976, Appl. 16 May 1975.
11. Belitz, H.-D.; Chen,W.; Jugel,H.; Treleano,R.; Wieser,H. unpublished results.
12. Ney,K.H. Z.Lebensm.Unters.Forsch., 1971, 147, 64.
13. Nozaki,Y.; Tanford,C. J.Biol.Chem., 1971, 246, 2211; Tanford,C.: J.Amer.Chem.Soc., 1962, 184, 4240
14. Wieser,H.; Belitz,H.-D. Z.Lebensm.Unters.Forsch. 1976, 160,383.
15. Ariyoshi,Y. Agr.Biol.Chem., 1976, 40, 983.
16. Brussel,L.B.P.; Peer,H.C.; v.d.Heijden,A. Z.Lebensm.Unters. Forsch., 1975, 159, 337.
17. Fujino,M.; Wakimasu,M.; Tanaka,K. Naturwissenschaften, 1973, 60, 351.
18. Lapidus,M.; Sweeney,M. J.Med.Chem., 1973, 16, 163.
19. Mazur,R.H.; Schlatter,J.M.; Goldkamp,A.H. J.Amer.Chem.Soc. 1969, 91, 2684.
20. Mazur,R.H.; Goldkamp,A.H.; James,P.A.; Schlatter,J.M. J.Med.Chem., 1970, 13, 1217.
21. Mazur,R.H.; Reuter,J.A.; Swiatek,K.A.; Schlatter,J.M. J.Med.Chem., 1973, 16, 1284.
22. Belitz,H.-D.; Wieser,H. Z.Lebensm.Unters.Forsch.,1976,160,251.
23. Treleano,R.; Belitz,H.-D.; Jugel,H.; Wieser,H. Z.Lebensm.Unters.Forsch. 1978, 167, 320.
24. Wieser,H.; Belitz,H.-D. Z.Lebensm.Unters.Forsch.,1975,159,65.
25. Pautet,F.; Nofre,C. Z.Lebensm.Unters.Forsch.,1978,166,167.

26. Gasteiger,J.; Marsili,M. Tetrahedron Letters, 1978, 3181.
27. Gasteiger,J.; Marsili,M. Proceed.IVth Internat.Conference on Computers in Chemical Research and Education, Novosibirsk, UdSSR, 1978.
28. Gasteiger,J.; Marsili,M. unbublished results.
29. Ueda, T.; Kobatake, Y. J.Membr.Biol., 1977, 34, 351.
30. Adams, W.J.; Geise,H.J.; Bartell,L.S. J.Amer.Chem.Soc., 1970, 92, 5013.
31. Anderson,J.E.; Glazer,E.S.; Griffith,D.L.; Knorr,R.; Roberts, J.D. J.Amer.Chem.Soc. 1969, 91, 1386.
32. Hendrickson, J.B. J.Amer.Chem.Soc. 1967, 89, 7036.
33. Hendrickson, J.B.; Boeckman,Jr.R.K.; Glickson,J.D.; Grundwald, E. J.Amer.Chem.Soc. 1973, 95, 494.
34. Lambert,J.B.; Papay,J.J.; Khan,S.A.; Kappauf,K.A.; Magyar, E.S. J.Amer.Chem.Soc. 1973, 95, 494.
35. Saunders, M. Tetrahedron, 1967, 23, 2105.

RECEIVED May 7, 1979.

5

Chemistry of Sweet Peptides

YASUO ARIYOSHI

Central Research Laboratories, Ajinomoto Co., Inc. 1-1, Suzuki-cho, Kawasaki-ku, Kawasaki, 210, Japan

It is known that sweet-tasting compounds are quite common and their chemical structures vary widely. In order to establish a structure-taste relationship, a large number of compounds have been tested, and several molecular theories of sweet taste have been proposed by different groups. At present, the phenomenon of sweet taste seems best explained by the tripartite functioning of the postulated AH, B (proton donor-acceptor) system and hydrophobic site X (1, 2, 3, 4, 5). Sweet-tasting compounds possess the AH-B-X system in the molecules, and the receptor site seems to be also a trifunctional unit similar to the AH-B-X system of the sweet compounds. Sweet taste results from interaction between the receptor site and the sweet unit of the compounds. Space-filling properties are also important as well as the charge and hydrophobic properties. The hydrophile-hydrophobe balance in a molecule seems to be another important factor.

After the finding of a sweet taste in L-Asp-L-Phe-OMe (aspartame) by Mazur *et al.* (6), a number of aspartyl dipeptide esters were synthesized by several groups in order to deduce structure-taste relationships, and to obtain potent sweet peptides. In the case of the peptides, the configuration and the conformation of the molecule are important in connection with the space-filling properties. The preferred conformations of amino acids can be shown by application of the extended Hückel theory calculation. However, projection of reasonable conformations for di- and tripeptide molecules is not easily accomplished.

In the course of investigations of aspartyl dipeptide esters, we had to draw their chemical structures in a unified formula. In an attempt to find a convenient method for predicting the sweet-tasting property of new peptides and, in particular, to elucidate more definite structure-taste relationships for aspartyl dipeptide esters, we previously applied the Fischer projection technique in drawing sweet molecules in a unified formula (4).

The sweet-tasting property of aspartyl dipeptide esters has been successfully explained on the basis of the general structures shown in Figure 1 (4). A peptide will taste sweet when it takes

0-8412-0526-4/79/47-115-133$05.00/0

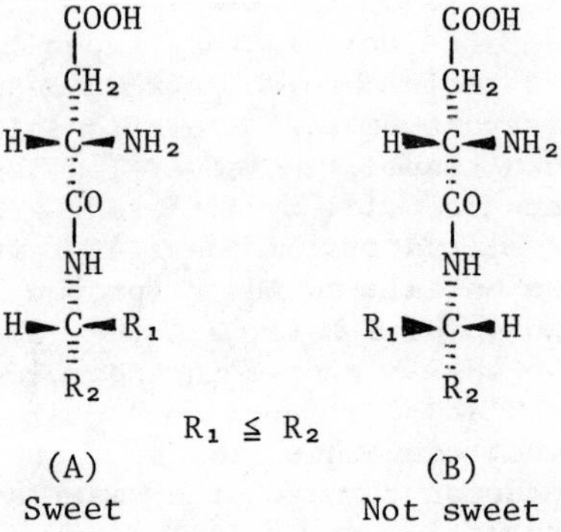

COOH
|
CH_2
H►C◄NH_2
CO
|
NH
H►C◄R_1
R_2

(A)
Sweet

COOH
|
CH_2
H►C◄NH_2
CO
|
NH
R_1►C◄H
R_2

(B)
Not sweet

$R_1 \leqq R_2$

Figure 1. General structure for sweet peptides: R_1 = small hydrophobic side chain (1 ~ 4 atoms); R_2 = larger hydrophobic side chain (3 ~6 atoms) (4)

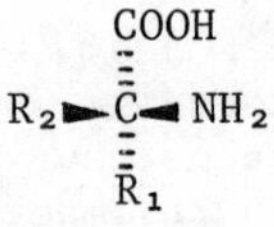

COOH
R_2►C◄NH_2
R_1

Figure 2. General structure for sweet amino acids: R_1 is not restricted; R_2 = H, CH_3, or C_2H_5 (12)

the formula (A), but not when it takes the formula (B), where R_2 is larger than R_1. R_1 is a small hydrophobic side chain with a chain length of 1∿4 atoms and R_2 is a larger hydrophobic side chain with a chain length of 3∿6 atoms. R_1 in formula (A) serves as the hydrophobic binding site (X). In the formula (A), when R_1 and R_2 are sufficiently dissimilar in size, the sweetness potency will be intense, whereas when R_1 and R_2 are of similar size, the potency will be weak (Table 1).

The structure-taste relationships will be discussed in detail. Dipeptide esters are closely related to amino acids in chemical structure and properties. Hence, we selected amino acids as the standard to which sweet peptides were related. The structural features of sweet-tasting amino acids have been best explained by Kaneko (12) as shown in Figure 2, in which an amino acid will taste sweet when R_2 is H, CH_3 or C_2H_5, whereas the size of R_1 is not restricted if the amino acid is soluble in water.

In the case of aspartyl dipeptide esters, proton donor AH is a free α-amino group, and proton acceptor B is a free β-carboxyl group. Therefore, the aspartyl part could be readily arranged to meet the structural requirements for sweet taste defined by Kaneko through the Fischer projection formula. The distance between the free α-amino and β-carboxyl groups was considered to be within the range defined for sweet molecules. In the case of the second amino acid such as Phe-OMe of aspartame, however, somewhat greater flexibility in drawing configurations was afforded by the interchange of atoms or groups attached to the asymmetric carbon atom of the second amino acid. This part of amino acid could be replaced by a great variety of L- or D-amino acid esters without losing the sweetness. This suggests that the sweet taste receptor site sees only the size and shape of this part, apart from the AH-B system of L-aspartic acid. It seems that the taste receptor sees the second amino acids as an alkyl side chain in the case of sweet amino acids.

In order to avoid confusion and to unify the system, the molecular structure was projected so that the largest side chain attached to the asymmetric carbon atom would be at the bottom of the formula and the amino group of the peptide bond in the upper position as shown in Figure 1. The remaining two groups such as a hydrogen atom and a smaller side chain, then laid in front of the projection plain. The orientation of the hydrogen atom and the smaller side chain depends on the configuration and the size of the two side chains of the amino acid ester. It was considered that the taste of the dipeptide esters changed according to the size and shape of the second amino acid. For instance, a sweet peptide, L-Asp-L-Phe-OMe (1), corresponds to the formula (A), where R_1 is a methyl ester group and R_2 is a benzyl group, whereas a nonsweet peptide, L-Asp-D-Phe-OMe (2), corresponds to the formula (B), where R_1 is a methyl ester group and R_2 is a benzyl group. This evidence suggests that a peptide will taste sweet when it takes the formula (A), but not when it takes the formula

Table 1. Taste of aspartyl peptides

Compounds	Projection formula*	R_1	R_2	Taste**	Lit.
1. L-Asp-L-Phe-OMe	A	$COOCH_3$	$CH_2C_6H_5$	180	6
2. L-Asp-D-Phe-OMe	B	CH_3OOC	$CH_2C_6H_5$	–	6
3. ε-Ac-D-Lys				5-10	4
4. L-Asp-Gly-OMe	A,B	H	$COOCH_3$	8	4
5. L-Asp-Gly-OEt	A,B	H	$COOCH_2CH_3$	13	7
6. L-Asp-Gly-OC_6H_{11}	A,B	H	$COOC_6H_{11}$	13	4
7. L-Asp-D-Ala-OMe	A	CH_3	$COOCH_3$	25	8
8. L-Asp-L-Ala-OMe	B	H_3C	$COOCH_3$	–	8
9. L-Asp-D-Abu-OMe	A	CH_2CH_3	$COOCH_3$	16	9
10. L-Asp-L-Abu-OMe	B	CH_3CH_2	$COOCH_3$	0	9
11. L-Asp-Gly-OPr^n	A,B	H	$COOCH_2CH_2CH_3$	14	4
12. L-Asp-D-Ala-OPr^n	A	CH_3	$COOCH_2CH_2CH_3$	170,125	8,9
13. L-Asp-D-Abu-OPr^n	A	CH_2CH_3	$COOCH_2CH_2CH_3$	95	9
14a. L-Asp-L-Nva-OMe	A	$COOCH_3$	$CH_2CH_2CH_3$	4	9
14b. L-Asp-L-Nva-OMe	B	$CH_3CH_2CH_2$	$COOCH_3$	0	9
15. L-Asp-L-Nva-OEt	B	$CH_3CH_2CH_2$	$COOCH_2CH_3$	–	9
16. L-Asp-D-Nva-OPr^n	A	CH_2CH_3	$COOCH_2CH_2CH_3$	45	4
17. L-Asp-L-Nle-OMe	A	$COOCH_3$	$CH_2CH_2CH_2CH_3$	45	9
18a. L-Asp-L-Nle-OEt	A	$COOCH_2CH_3$	$CH_2CH_2CH_2CH_3$	5	7
18b. L-Asp-L-Nle-OEt	B	$CH_3CH_2CH_2CH_2$	$COOCH_2CH_3$	0	7
19. L-Asp-L-Cap-OMe	A	$COOCH_3$	$CH_2CH_2CH_2CH_2CH_2CH_3$	47	7
20. L-Asp-L-Cap-OEt	A	$COOCH_2CH_3$	$CH_2CH_2CH_2CH_2CH_2CH_3$	–	7
21. L-Asp-L-Ser(Ac)-OMe	A	$COOCH_3$	CH_2OCOCH_3	10	9
22. L-Asp-L-Ser(Bt^i)-OMe	A	$COOCH_3$	$CH_2OCOCH(CH_3)_2$	50	9
23. L-Asp-L-Ser(Bt^i)-OEt	A	$COOCH_2CH_3$	$CH_2OCOCH(CH_3)_2$	2-3	9
24. L-Asp-L-Thr(Bt^i)-OMe	A	$COOCH_3$	$CH(CH_3)OCOCH(CH_3)_2$	–	9
25. L-Asp-L-αThr(Bt^i)-OMe	A	$COOCH_3$	$CH(CH_3)OCOCH(CH_3)_2$	–	9

Compounds	Projection formula*	R_1	R_2	Taste**	Lit.
26. L-Asp-L-Ile-OMe	A	$COOCH_3$	$CH(CH_3)CH_2CH_3$	-	6
27. L-Asp-L-Lys-OMe	A	$COOCH_3$	$CH_2CH_2CH_2CH_2NH_2$	-	9
28. L-Asp-L-Lys(Ac)-OMe	A	$COOCH_3$	$CH_2CH_2CH_2CH_2NHCOCH_3$	1.2	9
29. L-Asp-L-*t*-HyNle-OMe	A	$COOCH_3$	$CH(OH)CH_2CH_2CH_3$	7	9
30. L-Asp-L-*e*-HyNle-OMe	A	$COOCH_3$	$CH(OH)CH_2CH_2CH_3$	18	9
31. L-Asp-L-MPA	A	CH_3	$CH_2C_6H_5$	50	10
32. L-Asp-L-HMPA	A	CH_2OH	$CH_2C_6H_5$	1	10
33. L-Asp-D-Ser-OPr*n*	A	CH_2OH	$COOCH_2CH_2CH_3$	320	9
34. L-Asp-D-Ala-OBu*n*	A	CH_3	$COOCH_2CH_2CH_2CH_3$	50	7
35. L-Asp-D-Ser-OBu*n*	A	CH_2OH	$COOCH_2CH_2CH_2CH_3$	70	9
36. L-Asp-D-Thr-OPr*n*	A	$CH(CH_3)OH$	$COOCH_2CH_2CH_3$	150	9
37. L-Asp-D-*a*Thr-OPr*n*	A	$CH(CH_3)OH$	$COOCH_2CH_2CH_3$	40	9
38. L-Asp-Gly-Gly-OMe	A,B	H	$CONHCH_2COOCH_3$	0	7
39. L-Asp-D-Ala-Gly-OMe	A	CH_3	$CONHCH_2COOCH_3$	3	9
40. L-Asp-D-Abu-Gly-OMe	A	CH_2CH_3	$CONHCH_2COOCH_3$	1	7
41. L-Asp-D-Val-Gly-OMe	A	$CH(CH_3)_2$	$CONHCH_2COOCH_3$	0	7
42. L-Asp-L-Ama(OFn)-OMe	A	$COOCH_3$	COO-fenchyl	22000	11
43. L-Asp-D-Ama(OFn)-OMe	B	CH_3OOC	COO-fenchyl	0	11

*See Figure 1 for projection formula.

**Numbers represent the sweetness potency of the compound as a multiple of sucrose. In addition, 0 = tasteless, - = bitter.

(B). This also suggests that the AH-B concept represents only a first approximation in the case of peptides. Certainly, the AH-B system is required in the molecule. However, the structural characteristics of the second amino acid sometimes may completely mask any AH-B effect. To test the above hypothesis, we have synthesized a number of peptides with or without a sweet taste. The C-C bonding in R_2 has been replaced by ether, thioether, amide or ester bond without losing sweetness. R_1 is a small side chain such as a methyl, ethyl, isopropyl, or hydroxymethyl group, or an ester having a small substituent. The exact chemical nature of these groups is not crucial.

The studies on peptides began with a correlation between sweet amino acids and peptides. Since the projection formula of L-Asp-Gly-OMe (4) is similar in size and shape to that of ε-Ac-D-Lys (3) which is sweet, we predicted that L-Asp-Gly-OMe would taste sweet in spite of the bitter taste in the literature. Therefore, we synthesized the peptide and tasted it. As expected, it was sweet and its sweetness potency was almost equal to that of ε-Ac-D-Lys. Thus, the dipeptide could be correlated to the amino acid. Lengthening (5) or enlargement (6) of the alkyl group of the ester did not affect its sweetness potency (Table 1).

However, when a methyl group was introduced so as to protrude on the right of the projection formula of L-Asp-Gly-OMe, the resultant L-Asp-D-Ala-OMe (7) (8) was sweeter than L-Asp-Gly-OMe. This result suggests that the methyl group is involved in a hydrophobic interaction at the receptor site and causes the increased sweetness potency. On the other hand, when a methyl group was introduced so as to protrude on the left of the projection formula of L-Asp-Gly-OMe, the resultant L-Asp-L-Ala-OMe (8) (8) was not sweet but bitter. Loss of sweetness suggests that interaction with the receptor site may be blocked by the methyl group. This also supports the idea that a dipeptide will not taste sweet when it takes the formula (B), in which R_1 protrudes on the left. Introduction of an ethyl group instead of the methyl group so as to protrude on the right of the projection formula of L-Asp-Gly-OMe gave L-Asp-D-Abu-OMe (9), which was 16 times sweeter than sucrose. Introduction of an ethyl group on the opposite side gave L-Asp-L-Abu-OMe (10), which was devoid of sweetness. The low level of sweetness of (9), as compared with (7), may show that the population of the sweet formula (A) may significantly decrease because the two groups do not differ greatly in size.

This idea gained further support when a methyl group was introduced on the right of the projection formula of L-Asp-Gly-OPr*n* (11) to give L-Asp-D-Ala-OPr*n* (12), which was 125 times sweeter than sucrose. L-Asp-D-Ala-OPr*n* was about 9 times sweeter than L-Asp-Gly-OPr*n*. In the molecule of L-Asp-D-Ala-OPr*n*, the sizes of $CH_3(R_1)$ and $COOCH_2CH_2CH_3(R_2)$ are sufficiently dissimilar. An ethyl group was introduced instead of the methyl group to give L-Asp-D-Abu-OPr*n* (13), which was 95 times sweeter than sucrose and was less sweeter than L-Asp-D-Ala-OPr*n*. Lengthening the alkyl group of the ester of L-Asp-D-Abu-OMe increased the sweetness potency; L-Asp-D-Abu-OPr*n* was 6 times sweeter than L-Asp-D-Abu-OMe. This fact may show that the sweeter compound (13) takes predominantly the sweet formula (A), since the sizes of the two groups are sufficiently dissimilar.

More interesting is the case of L-Asp-L-Nva-OMe (14). Since the sizes of $COOCH_3$ and $CH_2CH_2CH_3$ are almost equal, both the sweet formula(A) and the nonsweet formula (B) could be drawn for the dipeptide, so that it could be predicted that the peptide would be slightly sweet. In fact, it was only 4 times sweeter than sucrose. Of course, replacement of the methyl group by an ethyl group resulted in a compound (L-Asp-L-Nva-OEt (15)) lacking in sweetness as expected from its projection formula. Some dipeptides containing D-norvaline were also sweet, when R_1 and R_2 matched the sweet formula (A), e.g., L-Asp-D-Nva-OPr*n* (16) was 45 times sweeter than sucrose.

Thus, it is plausible that a sweet response does not always depend on the configuration of the second amino acid but mainly depends on the size and shape of this amino acid ester.

L-Asp-L-Nle-OMe (17) was strongly sweet, but L-Asp-L-Nle-OEt (18) was only slightly sweet. These differences could be easily predicted from examination of their projection formulas. Both the sweet formula (A) and the nonsweet formula (B) could be drawn for the latter peptide.

L-Asp-L-Cap-OMe (19) was sweet, whereas the ethyl ester (20) was not sweet but bitter, though we could draw the sweet formula (A) to it. This may show that the increased hydrophobicity in the molecule changed the property of the sweet peptide to a bitter property, because it has been known that bitter-tasting compounds are composed of charge and hydrophobic properties. This also suggests that the hydrophile-hydrophobe balance in a sweet molecule is a very important factor.

An alkyl side chain of the second amino acid could be replaced by an ester group without losing the sweetness, e.g., L-Asp-L-Ser(Ac)-OMe (21), L-Asp-L-Ser(Bt*i*)-OMe (22) and L-Asp-L-Ser(Bt*i*)-OEt (23) were sweet. The replacement of the L-serine by L-threonine or by L-*allo*threonine resulted in bitter compounds (24, 25). These results matched that the introduction of a methyl group into a sweet peptide, L-Asp-L-Nva-OMe, resulted in a bitter substance (L-Asp-L-Ile-OMe (26)). The methyl group may block the interaction between the peptides and the sweet receptor.

From the above discussion, we have concluded that a hydrophobic binding site is necessary for a series of potent sweet peptides. Next, we examined how a hydrophilic group would affect the sweetness potency.

The introduction of an amino group to L-Asp-L-Nle-OMe (17) resulted in a bitter compound (L-Asp-L-Lys-OMe (27)) and blocking the amino group recovered the sweetness by some extent (28). The introduction of a hydroxyl group into a peptide with the L-L configuration (17, 31) resulted in a diminution in the potency

(29, 30, 32).

Contrary to the peptides with the L-L configuration, the introduction of a hydroxyl group into the L-D peptides did not always result in a diminution of their potencies, but sometimes increased their potencies. L-Asp-D-Ser-OR (R=Me, Et, Pr*n*, Pr*i*, Bu*n*, Bu*i* or *c*-hexyl) was sweeter than the corresponding peptides without a hydroxyl group, L-Asp-D-Ala-OR (9), e.g., compounds (33) and (35) were sweeter than compounds (12) and (34), respectively.

In the case of threonine-containing peptides, L-Asp-D-Thr-OR (R=Me or Pr*n*) was sweeter than L-Asp-D-Abu-OR which lacks a hydroxyl group of the D-threonine. On the contrary, when the D-threonine was replaced by D-*allo*threonine, the potency diminished significantly. The IR spectra of these peptides showed that the hydroxyl absorption had disappeared due to strong hydrogen bonding. It is considered that the apparent exceptions may be due to rigid intramolecular hydrogen bonding causing loss of hydrophilicity and allowing a hydrophobic group type interaction.

Returning to stereoisomerism, the relationships between stereoisomerism and taste will be discussed by using stereoisomers of aspartame (L-Asp-L-Phe-OMe) as model compounds. The lack of sweet taste in α-L-Asp-D-Phe-OMe (6) is readily explained after considering the projection formula ((2) in Figure 3), in which a small side chain on the left may cause elimination of sweetness. According to Mazur *et al*, β-D-Asp-L-Phe-OMe (6) is bitter though its projection formula ((44) in Figure 3) would suggest that it has a sweet taste. This result can not be explained fully. However, dipeptide esters carrying a small hydrophobic group on the 5th carbon from the carbon bearing the $AH(NH_2)$ often tasted bitter; e.g., L-Asp-L-Ile-OMe (26) and L-Asp-L-*a*Thr(Bt*i*)-OMe (25) were bitter. The location of $COOCH_3$ from the $AH(NH_2)$ in the peptide (44) corresponds to that of CH_3 in these peptides. The lack of sweet taste in β-L-aspartyl dipeptide esters such as β-L-Asp-Gly-OMe (7) and β-L-Asp-L-Phe-OMe (6) is readily explained after considering their projection formulas ((45) in Figure 3), in which the second amino acid lies on the left. This formula is incompatible with that defined for sweet amino acids, in which the second amino acid corresponds to R_2 of Figure 2. And also the peptide does not fit the spatial barrier model for the receptor site proposed by Shallenberger *et al*. (13). The lack of sweet taste in α-D-aspartyl dipeptide esters such as α-D-Asp-L-Phe-OMe (6) is interpreted analogously after considering their projection formulas ((46) in Figure 3).

Therefore, we have concluded that sweet-tasting aspartyl dipeptide esters can be drawn as the unified formula (A), whereas nonsweet peptides as (B) as shown in Figure 1.

There is no asymmetric carbon atom in aminomalonic acid molecule. When both of the carboxylic acids are substituted by esterification with different alcohols, optical isomers are generated. It is known that aminomalonic acid derivatives readily racemize in solution under ordinary conditions. L-Asp-Ama(OFn)-

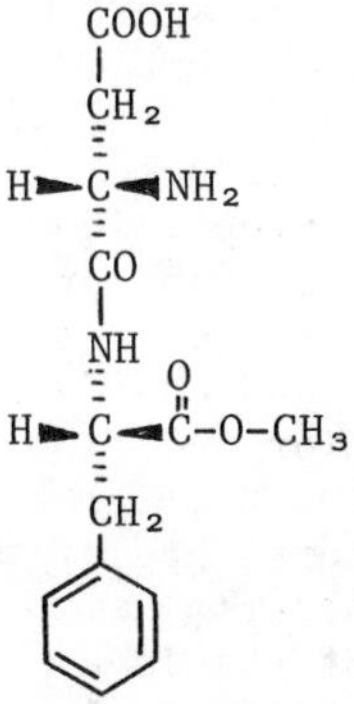

COOH
|
CH_2
H–C–NH_2
CO
|
NH
H–C–C(=O)–O–CH_3
CH_2
(phenyl)

180

α-L-L (1)

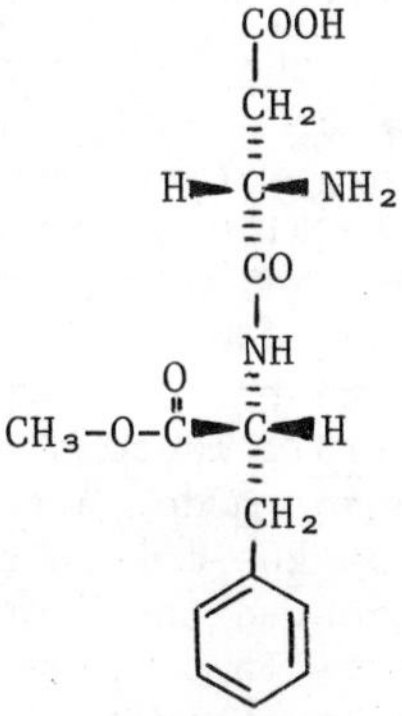

COOH
|
CH_2
H–C–NH_2
CO
|
NH
CH_3–O–C(=O)–C–H
CH_2
(phenyl)

–

α-L-D (2)

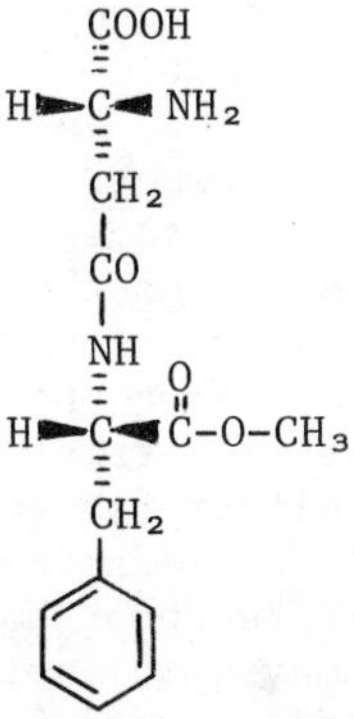

COOH
H–C–NH_2
CH_2
|
CO
|
NH
H–C–C(=O)–O–CH_3
CH_2
(phenyl)

–

β-D-L (44)

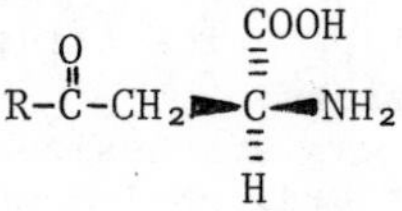

R–C(=O)–CH_2–C(COOH)(H)–NH_2

–

β-L-L(or D) (45)

R–C(=O)–C(H)(–CH_2–COOH)–NH_2

–

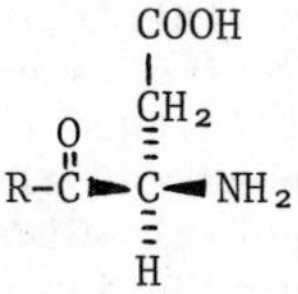

α-D-L(or D) (46)

R=Phe-OMe

Figure 3. Projection formulas of isomers of aspartame (L-Asp-L-Phe-OMe)

OMe was found by Fujino *et al.*(11) to be 22000∿33000 times sweeter than sucrose. It is not exactly known whether the sweet-tasting isomer has the L-L(or *S-R*) or the L-D(or *S-S*) configuration because of ready racemization. From the examination of its projection formula, it could be predicted that the L-L(or *S-R*) isomer (42), in which aminomalonic acid diester takes an L(or *R*)-configuration, would be sweet. This prediction agreed with that reported in the literature (14).

In Ama-L-Phe-OMe (47) (14, 15), it is also not known whether the sweet-tasting isomer has the L-L(or *S-S*) or the D-L(or *R-S*) configuration. In the case of aspartyl dipeptide esters, the L-L isomer was sweet. By analogy, other researchers deduced that the L-L(or *S-S*) isomer ((47b) in Figure 4) would be sweet. However, it seemed to us that the D(or *R*)-configuration would be preferred for the aminomalonic acid because the D-L(or *R-S*) isomer ((47a) in Figure 4) was compatible with the sweet formula and could also fit the spatial barrier model (13), whereas the L-L(or *S-S*) isomer could neither fit the receptor model nor meet the sweet formula.

Further examinations of the molecular features and of the model of receptor have suggested that several aspartyl tripeptide esters may also taste sweet. In confirmation of the idea, several tripeptide esters have been synthesized. In the first place, L-Asp-Gly-Gly-OMe (38) was synthesized as an arbitrarily-selected standard of tripeptides, because it was considered that this peptide ester had the simplest structure, and correlation of other peptides to (38) was easy. The tripeptide ester was predicted that it would be slightly sweet or tasteless because its projection formula was similar in size and shape to that of L-Asp-Gly-OBu^{n} which is 13 times sweeter than sucrose (16) and because it is more hydrophilic than the dipeptide. The tripeptide (38) was devoid of sweetness and almost tasteless.

Next, L-Asp-D-Ala-Gly-OMe (39) was synthesized in order to evaluate the contribution of a small side chain, which is properly oriented to elicit sweetness in the projection formula. The peptide was speculated to be sweet. As expected, it was sweet.

L-Asp-D-Abu-Gly-OMe (40) was selected as a next candidate in order to determine its sweetness intensity relative to L-Asp-D-Ala-Gly-OMe (39). The sweetness intensity of this peptide was predicted to be lower than that of L-Asp-D-Ala-Gly-OMe after examining their formulas. As expected, the synthesized L-Asp-D-Abu-Gly-OMe was sweet, and its sweetness intensity was lower than that of L-Asp-D-Ala-Gly-OMe.

Finally, L-Asp-D-Val-Gly-OMe (41) was synthesized in order to see whether it remained sweet. The peptide was devoid of sweetness and almost tasteless, though D-valine-containing aspartyl dipeptide esters such as L-Asp-D-Val-OPr^{n} (17) and L-Asp-D-Val-OPr^{i} (8, 17), which are similar to the tripeptide ester in size and shape and have potent sweet taste.

As mentioned above, the second amino acid of the sweet aspartyl dipeptide esters could be replaced by dipeptide esters such as D-Ala-Gly-OMe and D-Abu-Gly-OMe without losing the sweetness. However, their sweetness potencies were considerably lower than those of aspartyl dipeptide esters with the similar size and shape. Replacement of the second amino acid of a sweet aspartyl dipeptide ester such as L-Asp-Gly-OBu^n by Gly-Gly-OMe resulted in losing the sweetness ((38) in Figure 5), in spite of its similarity in the projection formula to that of the sweet dipeptide ester. These facts suggest that the tripeptide esters are more hydrophilic than the dipeptide esters and the hydrophilic property caused the sweetness intensity to decrease. The conformation of the tripeptide esters has, of course, influence on the elicitation of sweetness in connection with the space-filling properties of sweet compounds. However, the conformational problem can not be discussed here because it has not been investigated.

In the case of small-sized sweeteners such as glycine (48) and alanine (49), the sweetness sensation occurs only by the AH-B system and the sweetness intensity is low, as described previously. In the case of medium-sized sweeteners such as aspartyl dipeptide esters, two types of interaction have been considered. Among the aspartyl dipeptide esters without a hydrophobic binding site such as L-Asp-Gly-OPr^n (11), the sweetness sensation has occurred only by the AH-B system, like glycine, and the sweetness intensity is comparatively low. On the other hand, introduction of a small hydrophobic group into the sweet molecule so as to interact with a hydrophobic site of the receptor results in a sweeter compound such as L-Asp-D-Ala-OPr^n (12). The small hydrophobic group introduced plays a role in enhancing the sweetness intensity by forming a hydrophobic bond with the receptor site. This fact has been successfully explained by the theory of the AH-B-X system, in which X is the "dispersion" site proposed by Kier and has been proved experimentally by us to be a hydrophobic binding site (4). Therefore, in the case of medium-sized molecules, we have been able to conclude that formation of a hydrophobic bond causes the sweetness potency to increase. On the other hand, in the case of aspartyl tripeptide esters (39, 40), it appears that a small hydrophobic site for hydrophobic interaction is necessary to fit the receptor site.

One problem that remains is the mode of interaction between the sweet peptides and the receptor site. Despite a great number of studies, the mechanism of action of sweet stimuli on the receptor is not well known. Stereoisomerism can be responsible for differences in taste responses, and space-filling properties are also very important. These facts suggest that the receptor site exists in a three-dimensional structure. In this connection, the sense of sweet taste is subject to the "lock and key" of biological activity.

The above discussions, in conjunction with previous results, support our previous idea that the receptor site for sweet taste is composed of the AH-B-X system and its most likely shape is a "pocket" as shown in Figure 6 (5). In this model, the spatial

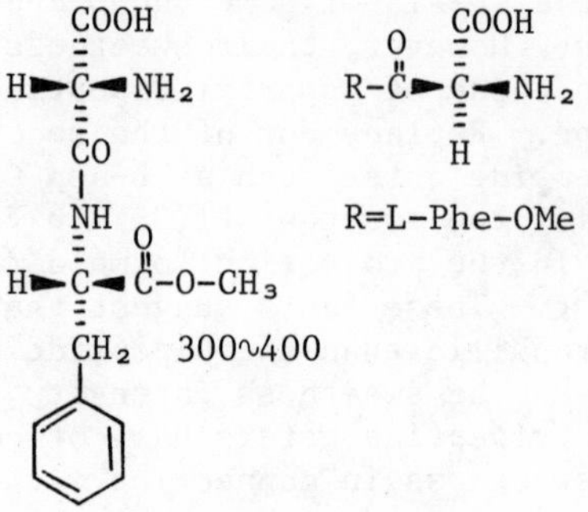

COOH
H►C◄NH₂ ...

D-L
Sweet formula
(47a)
300∿400

L-L
Nonsweet formula
(47b)
R=L-Phe-OMe

Figure 4. Projection formulas of isomers of Ama-L-Phe-OMe

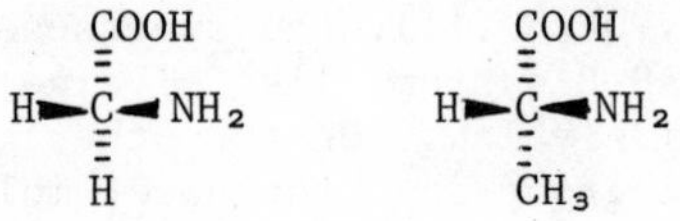

Gly (48) 0.9

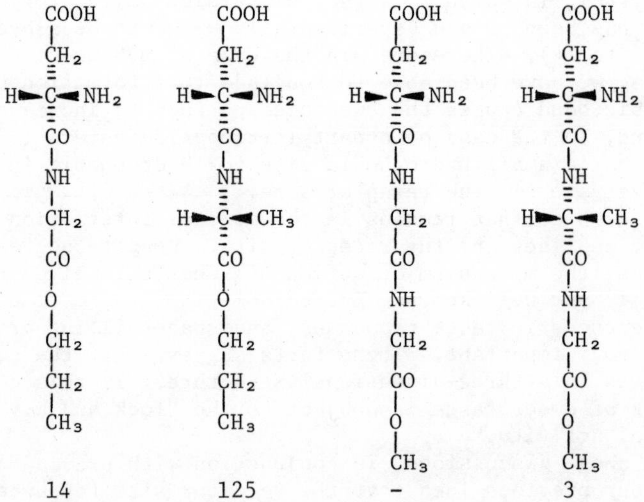

D-Ala (49) 1

(11) 14

(12) 125

(38) –

(39) 3

Figure 5. Projection formulas of various compounds

barriers are probably present at the back and on the right. The barrier at the back was found for amino acids (13) and that on the right for sulfamates (18). Interchange of 3'OH and 4'OMe groups on 8-desoxyphyllodulcin molecule has resulted in loss of sweetness (19). This may be explained by the presence of another spatial barrier being located on the left as described previously (5).

The conformational analyses of aspartame by means of NMR spectroscopy (20, 21, 22), though some discrepancies were found in these results, have suggested that the small side chain (R_1) is spatially directed to the front of the receptor model. The configurational change at the second asymmetric carbon in the sweet structure (A) gives the nonsweet structure (B). The loss of sweetness may be explained by the idea that interaction of (B) with the receptor site is interfered at the R_1 by the spatial barrier which may correspond to that found for amino acids. Consequently, the R_1 in the sweet structure (A) would be directed to the front, most likely to a little to the right of the front in the model. An examination of a model of 8-desoxyphyllodulcin also suggests that the hydrophobic binding site (X) is directed to this part. The sweet conformation for phyllodulcin (23) also may support this idea.

These considerations may lead to the conclusion that the receptor site (AH-B system) is surrounded by barriers from all sides, and a hydrophobic binding site (X) would be located on the right wall, a little to the front of the AH-B site. Therefore, it seems to us that the receptor site can be described as the "pocket" with the AH-B-X system inside it as shown in Figure 6. Figure 7 gives a schematic representation of the interaction between a sweet peptide, L-Asp-D-Ala-OPr^{n}, and the receptor site (24).

The receptor model seemed to be consistent with a variety of sweet compounds. An application to various sweet compounds will be discussed elsewhere. On the other hand, various types of the receptor model for sweet substances have been proposed by different groups (11, 13, 18, 22, 25, 26, 27).

Abbreviations follow the recommendation of the IUPAC-IUB Commision on Biochemical Nomenclature in J. Biol. Chem., 1966, 241, 2491; 1967, 242, 555; 1972, 247, 977. Other abbreviations used: Pr*n*, *n*-propyl; Pr*i*, *i*-propyl; Bu*n*, *n*-butyl; Bt*i*, *i*-butyroyl; Fe, fenchyl; Cap, capryline=α-amino-octanoic acid; *a*Thr, *allo*threonine; HyNle, β-hydroxynorleucine; MPA, methylphenethylamine, HMPA, hydroxymethylphenethylamine; Ama, aminomalonic acid.

Acknowledgement: The author wishes to thank to Dr. H. Wakamatsu and Dr. N.Sato for their interest in this work, and to Dr. T.Yamatani, Mr. N.Yasuda and Mr. F.Kakizaki for their helpful discussions and skillful experiment, and to Miss H. Furukawa and her group for the sweetness evaluations.

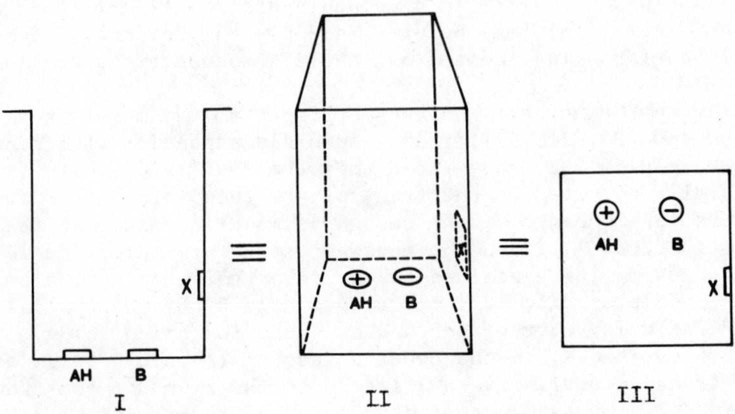

Figure 6. Schematic of the receptor site for sweet taste (5) (I is representation seen from the front, II from the upper front, and III from the top)

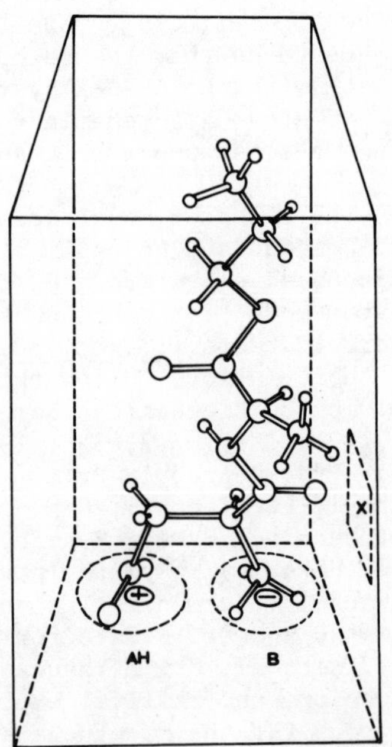

Figure 7. Schematic of the interaction between L-Asp-D-Ala-OPr^{n} and the receptor

Literature Cited

1. Shallenberger, R.S.; Acree, T.E. Nature, 1967, 216, 480.
2. Kier, L.B. J. Pharm. Sci., 1972, 61, 1394.
3. Birch, G.G.; Shallenberger, R.S. "Molecular Structure and Function of Food Carbohydrate", Birch, G.G.; Green, L.F. Ed., Applied Science Publishers, New York, 1973; p.9.
4. Ariyoshi, Y. Agric. Biol. Chem., 1976, 40, 983; see also Kagaku to Seibutsu, 1974, 12, 274.
5. Ariyoshi, Y. Kagaku Sosetsu (Chemistry of Taste and Smell), The Chemical Society of Japan Ed., Japan Scientific Societies Press, Tokyo, 1976, 14, 85.
6. Mazur, R.H.; Schlatter, J.M.; Goldkamp, A.H. J. Am. Chem. Soc., 1969, 91, 2684.
7. Ariyoshi, Y. Unpublished results; see also reference 5 for compounds (18a, 18b, 19 and 20). Compounds (38, 40 and 41) were presented briefly at the 5th International Congress of Food Science and Technology (Kyoto), September 1978.
8. Mazur, R.H.; Reuter, J.A.; Swiatek, K.A.; Schlatter, J.M. J. Med. Chem., 1973, 16, 1284.
9. Ariyoshi, Y.; Yasuda, N.; Yamatani, T. Bull. Chem. Soc. Jpn., 1974, 47, 326.
10. Mazur, R.H.; Goldkamp, A.H.; James, P.A.; Schlatter, J.M. J. Med. Chem., 1970, 13, 1217.
11. Fujino, M.; Wakimasu, M.; Mano, M.; Tanaka, K.; Nakajima, N.; Aoki, H. Chem. Pharm. Bull., 1976, 24, 2112.
12. Kaneko, T. J. Chem. Soc. Japan, 1939, 60, 531.
13. Shallenberger, R.S.; Acree, T.E.; Lee, C.Y. Nature, 1969, 221, 555.
14. Fujino, M.; Wakimasu, M.; Tanaka, K.; Aoki, H.; Nakajima, N. "Proceedings of the 11th Symposium on Peptide Chemistry", Kotake, H. Ed., Protein Research Foundation, Minoh-shi, Osaka, 1973; p.103.
15. Briggs, M.T.; Morley, J.S. Brit. Patent, 1,299,265 (1972).
16. Ariyoshi, Y.; see also reference 5.
17. Ariyoshi, Y. *et al.*, 100 times sweeter than sucrose; see also reference 5.
18. Pautet, F.; Nofre, C. Z. Lebensm. Unters.-Forsch., 1978, 166, 167.
19. Yamato, M.; Kitamura, T.; Hashigaki, K.; Kuwano, Y.; Yoshida, N.; Koyama, T. Yakugaku Zasshi, 1972, 92, 367.
20. Goodman, M.; Gilon, C. "Peptides 1974", Wolman Y. Ed., Halsted Press, New York, 1975; p.271.
21. Murai, A.; Ajisaka, K.; Nagashima, S.; Takeuchi, Y.; Kamisaku, M.; Kainosho, M. "Proceedings of the 13th Symposium on Peptide Chemistry", Yamada, S. Ed., Protein Research Foundation, Minoh-shi, Osaka, 1975; p.52.
22. Lelj, F.; Tancredi, T.; Temussi, P.A.; Toniolo, C. J. Am. Chem. Soc., 1976, 98, 6669.
23. DuBois, G.E.; Crosby, G.A.; Stephenson, R.A.; Wingard, Jr.R.E. J. Agric. Food Chem., 1977, 25, 763.

24. This was presented briefly at the 5th International Congress of Food Science and Technology (Kyoto), September 1978.
25. Horowitz, R.M.; Gentili, B. "Sweetness and Sweeteners", Birch,G.G.; Green,L.F.; Coulson, C.B. Ed., Applied Science Publishers Ltd., London, 1971; p.69.
26. Wieser, H; Jugel, H.; Belitz, H.-D. Z. Lebensm. Unters.-Forsch., 1977, 164, 277.
27. Temussi, P.A.; Lelj, F.; Tancredi, T. J. Med. Chem., 1978, 21, 1154.

RECEIVED August 2, 1979.

6

Bitterness of Peptides: Amino Acid Composition and Chain Length

KARL HEINZ NEY

Unilever Forschungsgesellschaft mbH Behringstrasse 154,
D-2000 Hamburg 50, West Germany

During our work on taste of foods we synthesized a series of peptides and soon came to the opinion, that the bitterness of peptides is caused by the hydrophobic action of amino acid side chains.

Here I think some remarks on hydrophobic interactions (1) would be appropriate. It is generally accepted now, that hydrophobic interactions are a contributing factor to protein behaviour and esp. to the formation of the secondary structure, e.g. helix. This means, that as shown in Figure 1 hydrophobic residues of the amino acids in a peptide are driven together by clusters of water molecules and so the secondary structure of a peptide or protein is formed. For the transfer from the helical to the stretched form, Tanford (2) found that the transfer free energy of the total protein results from the sum of the contributions of the single amino acid residues.

$$\Delta F = \sum \Delta f$$

The Δf values of the single amino acids given in Table I were determined by Tanford (2) from solubility data and they represent a measure of the hydrophobicity of an amino acid residue. Please note, that the values are relative to the methyl groups of glycine which is taken to be O. In Table II the taste of some "isomeric"-dipeptides is described. All the dipeptides are composed of the natural l-amino acids, as are all the examples, that will follow later. It is interesting to note, that the position of the amino acid has no influence on bitterness (3).

The value Q given represents the average hydrophobicity of a peptide and is obtained by summing the Δ f-values of the amino acid residues of a peptide and dividing by the number of the amino acid residues.

$$Q = \frac{\sum \Delta f}{n}$$

0-8412-0526-4/79/47-115-149$06.25/0

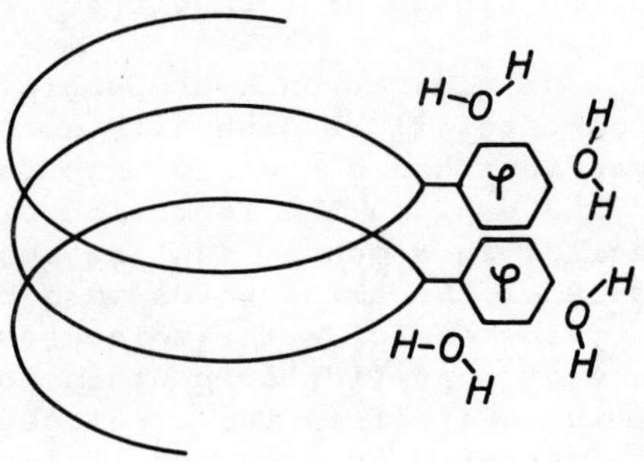

Figure 1. Hydrophobic interactions

Table I

Δf-values of the side chains of amino acids, representing their hydrophobicity, according to Tanford

Amino acid	Δ f-value cal/mol
Glycine	0
Serine	40
Threonine	440
Histidine	500
Aspartic acid	540
Glutamic acid	550
Arginine	730
Alanine	730
Methionine	1300
Lysine	1500
Valine	1690
Leucine	2420
Proline	2620
Phenylalanine	2650
Tyrosine	2870
Isoleucine	2970
Tryptophan	3000

Table II

Taste and Q-value of "Isomeric" dipeptides

Peptide	bitter	non-bitter	Q
Gly-Ala		x	365
Ala-Gly		x	365
Glu-Ala		x	640
Ala-Glu		x	640
Met-Ala		x	1015
Ala-Met		x	1015
Leu-Met	x		1860
Met-Leu	x		1860
Ala-Phe	x		1690
Phe-Ala	x		1690

You will have noticed in Table II, that the Q-values are much higher in the case of bitter dipeptides compared with the non-bitter dipeptides.

Table III shows a series of non-bitter dipeptides. It should be noted here that the Q-values are all below 1300. We can compare this with values of the following Table IV, which lists a series of bitter dipeptides with Q-values above 1400.

Table III

Q-values of further non-bitter dipeptides

Peptide	non-bitter	Q
Glu-Val	x	1120
Glu-Lys	x	1025
Gly-Gly	x	0
Gly-Asp	x	270
Ala-Asp	x	635
Ser-Asp	x	290
Ser-Glu	x	295
Val-Asp	x	1115
Val-Glu	x	1120
Ala-Ala	x	730
Asp-Asp	x	540
Glu-Asp	x	545
Glu-Gly	x	225
Gly-Ser	x	20
Gly-Thr	x	220
Val-Gly	x	845
Lys-Glu	x	1025

Table IV

Q-values of further bitter dipeptides

Peptide	bitter	Q
Leu-Tyr	x	2645
Leu-Leu	x	2420
Arg-Pro	x	1665
Asp-Phe	x	1595
Asp-Tyr	x	1705
Val-Leu	x	2055
Gly-Ile	x	1485
Gly-Phe	x	1325
Gly-Try	x	1500
Val-Val	x	1690
Glu-Phe	x	1600
Gly-Tyr	x	1435
Ala-Leu	x	1575

On Table V a series of bitter di- and tripeptides synthesized by Shiraishi (68) is given.

It follows therefore, that in the case of peptides from the natural l-amino acids no bitterness occurs when Q is below 1300, bitterness occuring only when the value Q exceeds 1400 (3).

Table V

Q-values of further bitter di- and tripeptides

Peptide	Q
Pro-Ala	1665
Ala-Pro	1665
Pro-Pro	2600
Val-Val	1690
Val-Pro	2145
Pro-Val	2145
Leu-Pro	2510
Pro-Leu	2510
Ile-Pro	2785
Pro-Ile	2785
Tyr-Pro	2735
Pro-Tyr	2735
Arg-Pro	1665
Lys-Pro	2050
Pro-Phe	2625
Phe-Pro	2625
Gly-Phe-Pro	1750
Phe-Pro-Gly	1750

If the Q-values lie between 1300 and 1400 no prediction can be made of the peptides bitterness.

It was interesting to see if our method can also be applied to individual l-amino acids. This means, that n = 1 and consequently in

$$Q = \frac{\sum \Delta f}{n}$$

Q equals Δ f.

As can be seen from Table VI, the individual l-amino acids also follow the rule. The only exceptions are lysine and proline, which have too high Q-values for non-bitter amino acids. However, a slight bitter note is detectable in the otherwise sweetish taste of lysine and proline.

In this context it is worth taking a brief look at the question of flavour enhancing qualities of glutamate, generally substances of the UMAMI-type as described by Shizuko Yamaguchi in her contribution to this symposium.

Kuninaka (4) proposed the following structural element for flavour intensifiers:

$$-\underset{\underset{O}{\|}}{C}-\underset{|}{\overset{|}{C}}-\underset{|}{\overset{|}{C}}-\underset{NH_2}{\overset{H}{C}}-COOH$$

but he pointed out, that the element is not absolute, as otherwise glutamine would have been a flavour enhancer.

Table VI

Q-values and taste of individual l-amino acids

l-Amino-Acid	bitter	non-bitter	Q
Glycine (opt. non active)		x	0
Serine		x	40
Threonine		x	440
Histidine		x	500
Aspartic acid		x	540
Glutamic acid		x	550
Arginine		x	730
Alanine		x	730
Methionine		x	1300
Lysine		x	1500
Valine	x		1690
Leucine	x		2420
Proline		x	2620
Phenylalanine	x		2650
Isoleucine	x		2970
Tryptophan	x		3000

Based on a series of examples from publications and patents, I would like to discuss, however, the hypothesis, that in order to achieve flavour enhancing, glutamate-like effect, a compound must have two negative charges. These should be located 3 to 9, preferably 4 to 6 C-atoms from one another. Instead of a C-atom, a S-atom can also occur. The presence of an α-amino group in l-configuration has additional flavour enhancing effect (5):

$$^{\ominus}OOC[-\overset{|}{\underset{|}{C}}-]_{n}COO^{\ominus} \qquad n = 1\text{-}7$$

The facts, on which our assumption is based, are given in Table VII.

An extension of our hypothesis to flavour-active nucleotides seems to be possible because these compounds also have negative charges at two different points of the molecule: in addition to the acidic phosphate group, they also possess a phenolic hydrogen.

It seems that the negative charges can also be on a peptide chain. Fujimaki describes the bitter masking action of peptides rich in glutamyl residues (29) and the isolation and identification of acidic oligopeptides from a flavour-intensifying fraction from fish protein hydrolysate (30).

Table VII

Facts on which our hypothesis is based

No.	Fact	Lit.
1)	Acc. to J. Solms only the dissociated form of l-glutamic acid is flavour-active	(6,92)
2)	l-Cystein-S-sulfonic acid has a similar effect to that of MSG	(7,8)
3)	l-Homocysteic acid has a similar effect to that of MSG	(9,10,11)
4)	l-Aspartic acid has a similar effect to that of MSG	(12)
5)	l-α-Aminoadipic acid has a similar effect to that of MSG	(11)
6)	Adipic acid makes the bitter after-taste of sweetners	(13,14,15)
7)	Succinic acid is comparable in its effect with that of MSG	(11,16)
8)	The flavour enhancing properties of the fruit acids - viz. malic acid, tartaric acid and citric acid - are known	(17,18,19,20)
9)	Lemon juice intensifies the flavour of strawberries	(21)
10)	The tastes of leguminose products are improved by treating with solutions of more than two of the following acids: malic acid, lactic acid, tartaric acid, citric acid	(22)
11)	The odour of garlic can be reduced by adding fumaric acid or maleic acid	(23)
12)	Glutathione (γ-glutamylcysteinyl-glycine) is reported to contribute towards the flavour of meat as an enhancer	(24)
13)	The diammonium salts of the dicarboxylic acids from malonic to sebacic acid are used as table salt-substitutes	(25)

Furthermore, it may be, that the well known action of polyphosphates in increasing the taste of chicken meat (31) or processed cheese (32) can be traced back on the negative charges of the polyphosphates.

Asparagine, unlike aspartic acid is completely lacking any flavour intensifying property, because one of the acidic groups was eliminated.

Also the findings on the derivatives of glutamic (33) acid are very interesting: if the glutamic acid is esterified or amidified, the flavour intensifying properties are lost.

I would like now to return to the topic of the Q-values dimensions. Since Tanford gave his Δ f-values in calories, the dimension of the Q-value is cal res^{-1}. All the Q-values mentioned in this paper are given in these dimensions.

Up to this point only amino acids, di- and tripeptides had been considered. However, we wanted to see if the Q-concept could be extended to higher peptides as well. Stepwise we synthesized a heptapeptide and followed the change of the taste. The following Table VIII shows this synthesis (5).

Table VIII

Peptide	bitter	non-bitter	Q
Glu-Lys		x	1025
Met-Glu-Lys		x	1116
Ala-Met-Glu-Lys		x	1020
Ile-Ala-Met-Glu-Lys	x		1410
Asp-Ile-Ala-Met-Glu-Lys		x	1265
Glu-Asp-Ile-Ala-Met-Glu-Lys		x	1163

As you can see, the di-, tri- and tetrapeptides have Q-values below 1300 and are not bitter. In the step leading to the pentapeptide the introduction of the strong hydrophobic isoleucine with its high Δf-value of 2970 confers a bitterness and correspondingly a Q-value of 1410. When aspartic acid with its low Δf-value of 540 is added,in the next step, the hexapeptide again becomes non-bitter with a Q-value of 1265. Glutamic acid - with a low Δ f-value of 550 - added in the final step gives a non-bitter heptapeptide with a Q-value of 1163. This example shows the influence of the amino acid residues as a polypeptide is synthesized and it gives a good demonstration of the possibilities of the method and we regarded it as a crucial experiment. Whereas in this example the bitter taste during the synthesis of peptides was followed, Table IX gives according to Minamiura (34) the degradation of a bitter peptide obtained from the action of Bacillus subtilis on casein.

Table IX

Degradation of a bitter peptide obtained from the action of Bacillus subtilis on Casein

Peptide	Q
Arg-Gly-Pro-Pro-Phe-Ile-Val	1891
Gly-Pro-Pro-Phe-Ile-Val	2085
Arg-Gly-Pro-Pro-Phe	1716
Gly-Pro-Pro-Phe	1963

We now wanted to extend the range to peptides of longer chain length. As you see from Table X, the Q-method works well up to eikosapeptides.

Table X

Q-values and taste of tri- to eikosapeptides

Peptide	bitter	non-bitter	Q
Val-Val-Val	x		1690
Ala-Ser-Phe		x	1140
Val-Val-Glu		x	1310
Pro-Gly-Gly-Glu		x	787
Ser-Pro-Pro-Pro-Gly	x		1508
Gly-Pro-Phe-Pro-Val-Ile	x		2085
Val-Ser-Glu-Glu-Glu-Asp-Ile-Ala-Met-Glu-Lys		x	815
Lys-Asp-Glu-Glu-Glu-Glu-Val-Glu-Ser-Gly-Pro-Asp-Ala-Pro-Leu-Pro-Ala-Glu		x	1121
Phe-Phe-Val-Ala-Pro-Phe-Pro-Glu-Val-Phe-Glu-Lys-Phe-Ala-Leu-Pro-Glu-Tyr-Leu-Lys	x		1912

Kauffmann and Kossel (35) isolated a series of oligopeptides from spinach and these are shown in Table XI.

Table XI

Q-values of non-bitter oligopeptides from spinach

Peptide	Q
Glu-Gly	225
Glu-(Gly,Ser)	196
Gly-(Glu,Ser)	196
Ala-(Glu,Gly-Ser)	330
Glu-(Gly,Gly,Ala)	320
Asp-(Glu,Gly,Ser,Ser)	234
Ser-(Gly,Gly,Thr)	120
Ala-(Glu,Glu,Gly,Ser)	374

As you see, the Q-values are extremely low and therefore the peptides non bitter.

As given in Table XII the Q-method was also successfully applied in the case of bitter peptides from the rennet-sensitive sequence of K-casein (36).

We published the Q-hypothesis in 1971 (3) and thus established for the first time a quantitative relationship between the amino acid composition of a peptide and its bitterness, as we introduced the Tanford values and so opened the way for a calculation of bitterness.

Table XII

Bitter peptides synthesized acc. to the rennet-sensitive sequence of K-Casein

Peptide	Q
Ser-Leu-Phe-Met-Ala	1428
Lys-His-Pro-Pro-His-Leu-Ser-Phe	1726
Lys-His-Pro-Pro-His-Leu-Ser-Phe-Met-Ala-Ile-Pro-Pro-Lys-Lys	2001

In Table XIII we have collated other former postulates for bitterness of peptides. The results are in agreement with the Q-rule, for example the sequence Gly-Pro-Pro-Phe postulated by Minamiura (34) to be the core of the bitterness has a high Q-value of 1963.

Table XIII

Former postulated requirements for bitterness of peptides

Amino acid or sequence inducing bitterness	Lit.	Q
-Leucine-	37,38,39	2420
-Try-Phe-Leu-	40	2647
-Gly-Pro-Pro-Phe-	34	1963
-2 neutral amino acids with large alkyl groups C ≥ 3		high
-1 neutral amino acid with a large alkyl group		high
C ≥ 3 with a short alkyl group	41	high
-1 neutral amino acid + 1 aromatic amino acid		high
-1 neutral amino acid + 1 basic amino acid		open

The same holds for the sequence Tyr-Phe-Leu, postulated by Fujimaki (40) to be essential for bitterness, here the Q-value is 2647. Also leucine, postulated earlier by Fujimaki (37,38,39) to be essential for bitterness, has a Δ f-value of 2420 and therefore contributes considerably to the Q-value of any peptide of which it forms a part.

Also the postulates of Kirimura (41) correspond to our theory.

It follows that the Q-concept represents a general rule for predicting bitterness under which the previously cited postulates are special cases.

The dipeptide glutamyl-tyrosine is bitter below pH 10,

and not bitter above pH 10. This coincides with the dissociation of the phenolic hydroxyl group of tyrosine. The corresponding dipeptide glutamyl-phenylalanine has no phenolic group, and is bitter over the whole pH-range. Q of this compound is 1660 (42).

The Q-concept has been assessed and accepted by the scientists (43-61) working in this field.

Series of bitter peptides have been isolated from enzymatic hydrolysates of proteins, esp. casein and soybean protein.

Figure 2 gives the sequence (61,62,63) of α_{s1}- casein - which represents about 40 % of casein - and shows the bitter peptides, that have been isolated. According to Mercier (63) the polypeptide chain of α_{s1}- casein contains 3 hydrophobic regions, viz. 1-44, 90-113 and 132-199. It is very interesting that all bitter peptides derived from α_{s1}- casein and isolated by the groups of Mercier (63), Matoba (65), Belitz (66), Solms (47), Hill (67) are located in these hydrophobic regions and have Q-values above 1400.

Figure 3 gives the sequence of β-casein - which represents 30 % of casein - and the bitter peptides derived from it and isolated by the groups of Clegg (49), Klostermeyer (46), Gordon (64). Here also the Q-values of the bitter peptides are above 1400. Please note, that no special single amino acid or sequence is needed to impart the bitter taste.

From soybean protein hydrolysates several series of bitter peptides have been isolated. As an example Table XIV shows bitter peptides isolated by Fujimaki (69,70). As before the high Q-values are evident.

Table XIV

Bitter peptides from peptic soya protein hydrolysates

Peptide	Q
Leu-Phe	2535
Leu-Lys	1960
Arg-Leu	1575
Arg-Leu-Leu	1856
Phe-Ile-Ile-Glu-Gly-Val	1766

From peptic Zein hydrolysates, Wieser and Belitz (71) isolated bitter peptides which are given in Table XV together with the corresponding high Q-values.

Regarding the whole picture of enzymatic hydrolysates we came to the conclusion, that certain proteins are more prone to yield bitter peptides than others. Therefore we tried to transfer our method also to proteins as well. This would enable a prediction to be made as to whether in the

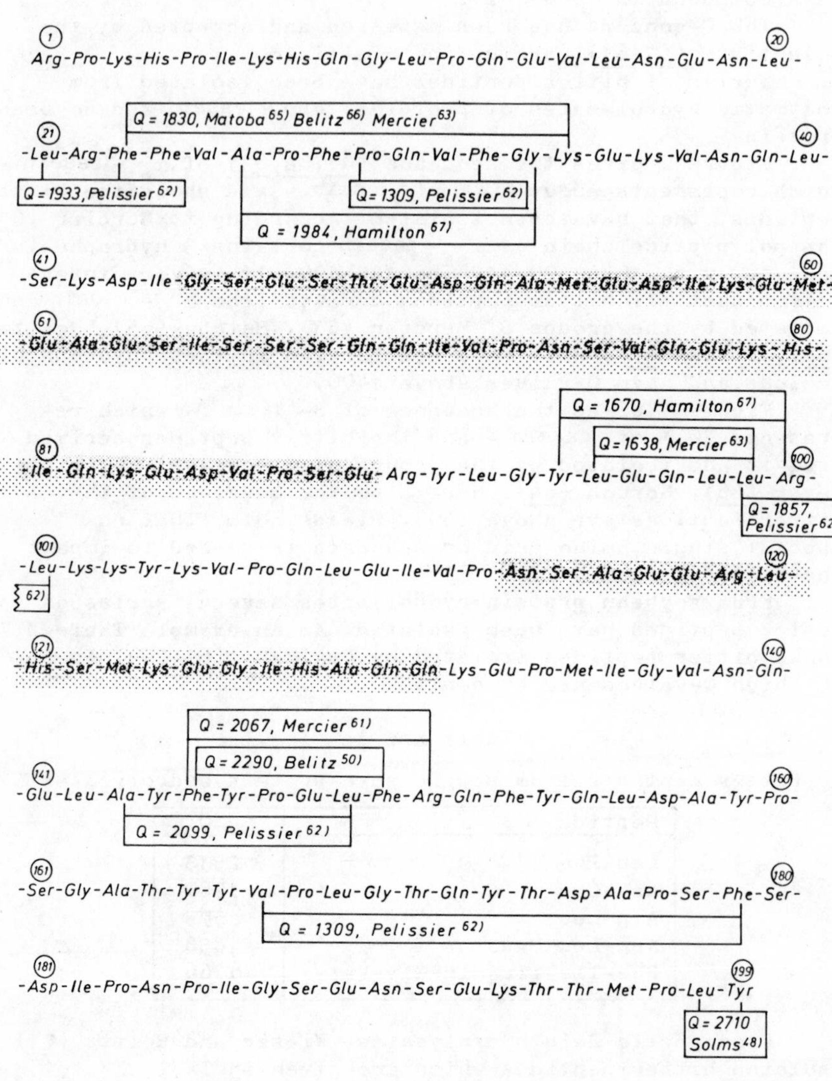

Figure 2. Bitter peptides from α_{S1}-casein (shaded) = hydrophilic regions

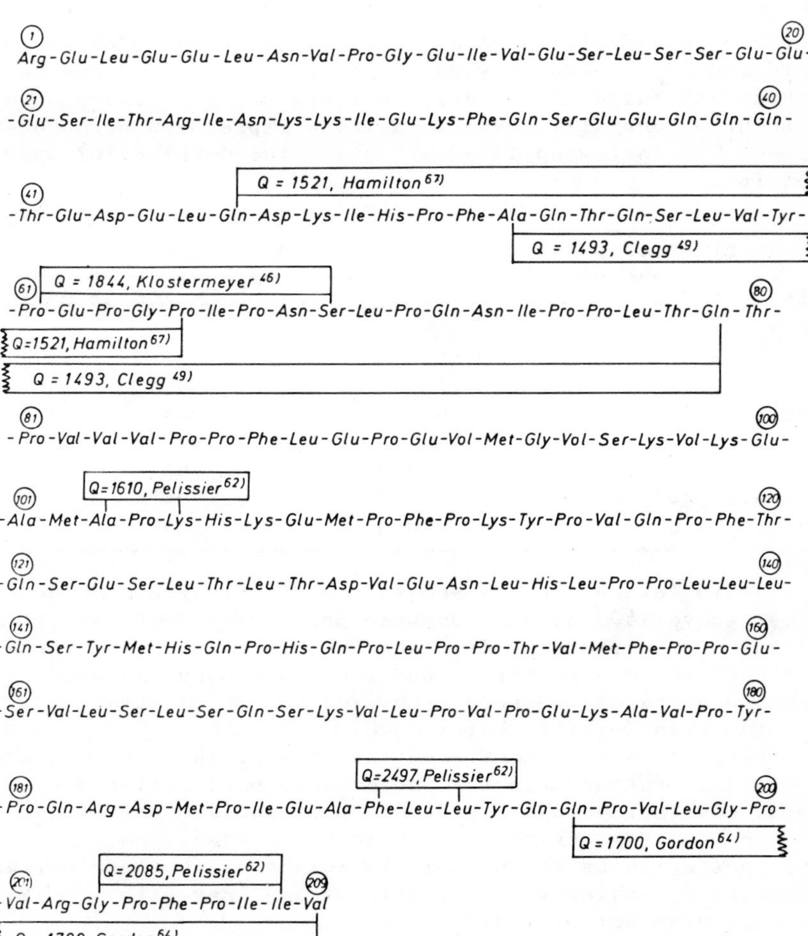

Figure 3. Bitter peptides from β-casein

Table XV

Bitter peptides from peptic Zein hydrolysates

Peptide	Q
Ala-Ile-Ala	1477
Ala-Ala-Leu	1293
Leu-Gln-Leu	1613
Leu-Glu-Leu	1797
Leu-Val-Leu	2177
Leu-Pro-Phe-Asn-Glu-Leu	1682
Leu-Pro-Phe-Ser-Glu-Leu	1688

course of a hydrolysis of a protein, bitter peptides would be formed (72). Generally pure proteins are considered to be without any taste. Secondary, tertiary and quaternary structures generally prevent a taste impression being obtained. The following Table XVI gives the Q-values of some proteins.

Table XVI

Q-values of proteins and bitter hydrolysates derived

Protein	Q	Bitter hydrolysates known
Collagen	1280	no
Gelatin	1280	no
Bovine muscular tissue	1300	no
Wheat Gluten	1420	yes
Zein	1480	yes
Soybean protein	1540	yes
Potato protein	1567	yes
Casein	1600	yes

It is interesting to see that proteins with high Q-values above 1400 as e.g. soybean protein, casein wheat gluten, potato protein, Zein are the "parents" of bitter peptides, whereas no bitter peptides have been isolated from hydrolysates prepared from collagen or gelatin, proteins with Q-values below 1300.

Petrischek (74) confirmed that the protein and not the protease is responsible for the occurence of bitter peptides. However, when the "parent" proteins are not bitter but the peptides derived from them are bitter, the questions arise as to why this is so and as to where we must place the molecular weight limits of peptides with Q $\gg$ 1400 that are also not bitter.

An indication of the values to be expected can be obtained from the results of our synthesis of bitter peptides with Q $>$ 1400 and molecular weights up to 2000 Dalton. Fujimaki (75) isolated from the peptic hydrolysate of soybean protein a non-dialysable bitter peptide of a molecular weight of about 2800 Dalton. Pilnik (76) found by the proteolysis of soybean protein in a membrane-filtration apparatus that no bitter peptides existed with molecular weights above 6000 Dalton. Clegg (49) obtained from digests of Casein with Papain a bitter peptide having a molecular weight of about 3000 Dalton.

Fujimaki (77,78) condensed bitter soybean protein hydrolysates in a Plastein-Reaction (79) and obtained non-bitter protein-like products, unfortunately without determination of molecular weights.

We studied the influence of chain length on the bitterness of peptides by gel permeation chromatography of enzymatic protein hydrolysates (80).

Table XVII sums up the results of these experiments. We can conclude, that a limit of about 6000 Dalton can be placed on the molecular weight.

Table XVII

Molecular weights and tastes of enzymatic hydrolysates

Parent Protein	Q	Molecular weight of hydrolysate in Dalton	Taste bitter	Taste non-bitter
Soybean protein	1540	4000	x	
Soybean protein	1540	12500		x
Casein	1605	4000	x	
Casein	1605	8000		x
Wheat Gluten	1420	5000	x	
Potato protein	1567	400	x	
Potato protein	1567	8000		x
Gelatine	1280	3000		x

Above this molecular weight, also peptides with a Q-value above 1400 will no longer exhibit bitter taste. It is clear therefore, that 2 ways exist to come to non-bitter protein hydrolysates. As demonstrated in Figure 4

a) choice of the starting material, this means proteins with Q-values below 1300

b) choice of the working conditions, this means, if the Q-value of the starting protein is above 1400, careful hydrolysis to obtain peptides with main molecular weights of above 6000 Dalton.

It should be pointed out, that we were concerned with presence or absence of bitterness. Bitterness in terms of sensory threshold values or bitterness ratings was not assessed.

What is now the current state of affairs of the Q-rule? As mentioned it has been accepted and applied by the scientists working in this field. The most comprehensive and careful assessment of the Q-rule has been carried out by Guigoz and Solms (54). They found that the rule can be applied to the majority of the bitter peptides known and observed, that only peptides containing glycine sometime do not comply fully with the rule. They therefore propose, that glycine should be left out of the calculations, which then gives Q-values higher than 1400 for all bitter peptides. Guigoz and Solms conclude that the Q-values should be a useful assessment of the relationship between amino acid composition and the bitter taste of peptides. Wieser and Belitz (81) have suggested a very interesting extension of the rule. They obtained the bitterness threshold values of

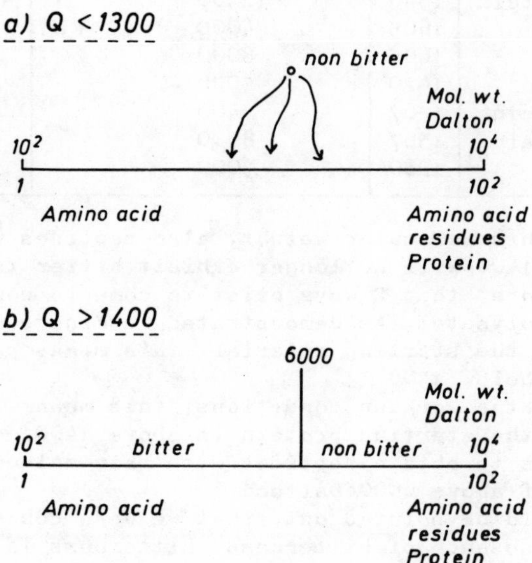

Figure 4. Molecular weight, average hydrophobicity Q, and bitter taste of peptides

di- and tripeptides by calculating the sum of the hydrophobicity of the "backbone" peptide consisting only of glycine residues adding to it the hydrophobicities of the side chains. In this way an estimation of the threshold values of di- and tripeptides was obtained.

We now investigated the hypothesis if the bitterness of lipids - and carbohydrates - could also be linked to hydrophobic interactions (82,83). Let us look first at the questions of hydroxylated fatty acids.

By the action of lipoxygenase and peroxydase on linolenic acid Grosch (84) obtained an intensively bitter tasting trihydroxyoctadecenoic acid. On the other hand, it is well known that monohydroxystearic acid and dihydroxystearic acid do not exhibit a bitter taste. We know from our studies on proteins that hydrophobicity plays a key role in determining bitterness. Lipids are, however, too hydrophobic to be bitter and bitterness here increases with increasing hydrophilicity. As a criterion for this diminution of hydrophobicity we applied the ratio of the number of carbon atoms of a molecule to the number of its hydroxyl groups. So the value $\frac{n_C}{n_{OH}}$ is obtained, which we have called the "R-value".

In Table XVIII the R-values for the hydroxylated C_{18} fatty acids are given.

Table XVIII

Bitterness of Hydroxy acids

Substance	n_C	n_{OH}	bitter	non-bitter	sweet	$\frac{n_C}{n_{OH}}=R$
Monohydroxystearic acid	18	1		x		18.00
Dihydroxystearic acid	18	2		x		9.00
Trihydroxyoctadecenoic acid	18	3	x			6.00

As the number of the hydroxyl groups changes, it is evident, that the accumulation of the 3 hydroxyl groups induces a bitterness: it can be seen that the R-value of the bitter substance is 6.00.

Wieske and Guhr investigated the taste properties of monoglycerides, diglycerides and phosphatides. We refer here to their findings (85).

As can be seen from Table XIX the R-values of bitter mono- and diglycerides are below 7.00.

In the case of phosphatides, we have made the assumption that one phosphatidyl-choline is equivalent to 2 hydroxyl groups. The following Table XX gives the results of phosphatides.

Table XIX
Bitterness of mono- and diglycerides

Substance	n_C	n_{OH}	bitter	non-bitter	sweet	$\frac{n_C}{n_{OH}}$ =R
Monobutyrin	7	2	x			3.50
Monocaprin	13	2	x			6.50
Monolaurin	15	2		x		7.50
Monomyristin	17	2		x		8.50
Monoglyceride of linseed oil	20	2		x		10.00
1,3 Dicaprylin	19	1		x		19.00
Tetraglycerolmono-caprylate	20	5	x			4.00
Tetraglycerolmono-laurate	24	5	x			4.80

Table XX
Bitterness of phosphatides

Substance	n_C	n_{OH}	bitter	non-bitter	sweet	$\frac{n_C}{n_{OH}}$ =R
1,2 Dicaprinoylphos-phatidylcholine	28	2		x		14.00
1,2 Dilauroylphos-phatidylcholine	32	2		x		16.00
Lyso-Laurophosphati-dylcholine	20	3	x			6.66
Lyso-Oleylphosphati-dylcholine	26	3		x		8.25

Similarly, we can see here, that above R = 7.00 no bitterness occurs.

Bitterness in terms of sensory threshold values or bitterness ratings was not assessed.

Having previously considered the glycerides we then studied glycerol itself and related compounds.

The following Table XXI gives the results. Here we found an interesting fact that - outside the fat area - with R = 1 sweet taste occurred and we therefore included sugars and derivatives in our considerations. The occurrence of monofunctional substituents like the hydroxyl groups always raises the question of stereochemistry, if the substituents are different. This question was not taken into consideration for the moment.

Table XXI

Bitterness of glycerol and derivatives

Substance	n_C	n_{OH}	bitter	non-bitter	sweet	$\frac{n_C}{n_{OH}}$ =R
Ethylene glycol	2	2			x	1.00
Glycerol	3	3			x	1.00
2,3 Dihydroxypropionic acid ethyl ester	5	2	x			2.50

According to Birch and Lindley (86) the sweetness of sugars decreases with increasing molecular weight. We therefore considered only mono- and disaccharides. Table XXII gives the results.

Table XXII

Bitterness of sugars and derivatives

Substance	n_C	n_{OH}	bitter	non-bitter	sweet	$\frac{n_C}{n_{OH}}$ =R
Glukose	6	5			x	1.20
Galaktose	6	5			x	1.20
Fruktose	6	5			x	1.20
Tetramethylglucose	6	1	x			6.00
Lactose	12	8			x	1.50
Saccharose	12	8			x	1.50
Cellobiose	12	8			x	1.50
Maltose	12	8			x	1.50
Trehalose	12	8			x	1.50
Arabinose	5	4			x	1.25
Xylose	5	4			x	1.25
Ribose	5	4			x	1.25
Desoxyribose	5	3			x	1.66
Methylglucopyranose	7	4			x	1.75
Athylglucopyranose	8	4	x			2.00
Propylglucopyranose	9	4	x			2.25
Butylglucopyranose	10	4	x			2.50
Phenylglucopyranose	12	4	x			3.00
Benzylglucopyranose	14	4	x			2.50
Inositol	6	6			x	1.00
Xylitol	5	5			x	1.00

As can be seen from the table, a sweet taste occurs when R has a value between 1.00 and 1.99; bitter compounds having R-values between 2.00 and 6.99. This is in full agreement with the finding of Birch and Lee (87) that reactions,which increase the hydrophobicity of sugars, generally lead to bitter products.

Bitterness of terpenoids, of purines like coffein, and of glucosides (88,89) may also be derived from hydrophobic interactions. See also the contribution of Belitz to this symposium.

A complete different mechanism seems to be present in the bitterness of salts , as two bitter sensations are differentiated (90): bitter I as elicited by stimuli like l-tryptophan, this would correspond to our "hydrophobic bitterness" and bitter II, elicited e.g. by $MgSO_4$. This bitter II seems to be triggered by ions. Kionka and Strätz (91) comparing 1 n solutions of the different alkali halogenides made a separation in three groups as shown in Table XXIII: salty, salty + bitter, bitter.

Table XXIII

Bitterness of salts

a) salty taste dominates. NaCl, KCl, LiCl, RbCl, NaBr, LiBr, NaJ, LiJ
b) salty and bitter: KBr
c) bitter dominates: CsCl, RbBr, CsBr, KJ, RbJ, CsJ

We give in Table XXIV the salts ordered in increasing sum of the ionic diameter and compare also the solubility in water. As can be derived from the Table, there is no relationship between the solubility of the salts in water and the taste. Molecular weights show a certain parallel: with increasing molecular weight the salts became bitter. An exception is KBr, which is bitter and salty, but according to the molecular weight should only be salty. A clear relation, however, exists between the sum of the ionic diameter of a salt and its bitterness. From LiCl with 4.98 Å to RbCl with 6.56 Å the salty taste dominates. KBr with 6.58 Å is salty and bitter and from RbBr with 6.86 Å to CsI with 7.74 Å the bitter taste dominates. It should be mentioned that $MgCl_2$ with 8.50 Å is also bitter,the same holds for $MgSO_4$.

Table XXIV

Relations between the bitterness of salts and their ionic diameter

Salt	Sum of the ionic diameter (Å)	Taste bitter	Taste salty	Solubility (g/100ml H_2O)	Molecular weight
LiCl	4.98		+	63.7	42.39
LiBr	5.28		+	145.0	86.85
NaCl	5.56		+	35.7	58.44
LiJ	5.76		+	151.0	133.84
NaBr	5.86		+	116.0	102.90
KCl	6.28		+	34.7	74.56
NaJ	6.34		+	184.0	148.89
RbCl	6.56		+	77.0	120.92
KBr	6.58	+	+	53.5	119.01
RbBr	6.86	+		98.0	165.38
CsCl	6.96	+		162.2	168.36
KJ	7.06	+		127.5	166.01
CsBr	7.26	+		124.3	212.81
RdJ	7.34	+		152.0	212.37
CsJ	7.74	+		44.0	259.81

Summary

Bitterness of a peptide is caused by the hydrophobic action of its amino acid side chains. By summing the hydrophobicities of the amino acid side chains of a peptide and dividing by the number of the amino acid residues, an average hydrophobicity Q is obtained. Peptides with Q-values below 1300 are not bitter, whereas peptides with Q-values higher than 1400 are bitter. This principle is valid for molecular weights up to approximately 6000 Dalton, above this limit peptides with Q 1400 are also not bitter.

Practically all known peptides with defined amino acid composition, chain length and flavour, whether isolated or synthetic, follow this principle and they number about 200 in 1978. It is therefore possible to predict the bitterness of any new peptide simply from its amino acid composition and chain length. Furthermore the danger of obtaining bitter peptides from enzymatic hydrolysis of a protein can also be predicted. For example, casein and soy protein, having high Q-values, are prone to produce bitter peptides on enzymatic hydrolysis, whereas collagen having a low Q-value does not give bitter peptides.

Hydrophobic interactions can also be used to provide information on the bitterness of lipids.

Bitterness of salts seems to be triggered by another mechanism.

Acknowledgements

I would like to thank

- Dr. Garg, for dicussions on hydrophobic interactions
- Dr. Polzhofer, for peptide syntheses
- Dr. Thomas, Dr. Fox and Mr. Oliver, for help in the English formulation
- Dr. Wirotama, for enzymatic hydrolysis
- Mrs. Guhr and Mr. Wieske, for experimental data
- Dr. Unbehend, for gel permeation chromatography
- Dr. Todt and Dr. Heider, for discussions on phosphatides
- Mrs. Müller and Miss Liebherr, for technical assistance.

Literature cited

1) Nemethy, G., Angew. Chemie 79, 260 (1967).
2) Tanford, C., J.Am.Chem.Soc. 84, 4240 (1960).
3) Ney, K.H., Z.Unters. Lebensm. Forsch. 147, 64 (1971).
4) Kuninaka, A., Symposium on Foods, The Chemistry and Physiology of Flavours, P. 515 ff., Westport/Conn. AVI 1967.
5) Ney, K.H., Z. Unters. Lebensm. Forsch. 146, 141 (1971).
6) Solms, J., Aroma- und Geschmacksstoffe in Lebensmitteln, P. 199 ff., Zürich, Forster 1967.
7) Sakakibara, E.A., Hagihara, F.N., A.P. 3214276 of 26/10/1965.
8) D.A.S. 1 492 729 of 25.9.1969.
9) Ajinomoto, Belg. P. 918 640 of 13.2.1963.
10) Kaheko, T., Toi, B., Ikeda, S., A.P. 3 259 505 of 5.7.1966.
11) Toi, B., Maeda, S., Ikeda, S., Furukawa, H., A.P. 3 109 741 of 5.11.1963.
12) Matsuda, AC., Shiba, A.M., D.A.S. 1 517 060 of 14.5.1969.
13) General Foods, A.P. 2 971 848 of 14.2.1961.
14) General Foods, Can.P. 663 877 of 12.5.1958.
15) General Foods, Brit. P. 855 892 of 14.2.1961.
16) Kirk-Othmer, 2. Ed. Vol. 2, P. 198, New York - London - Sidney; Intersc. Pupl. 1963.
17) Rudy, H., Fruchtsäuren, pag. 198, Heidelberg,Hüthig 1967.
18) Cumberland Pack. Co., A.P. 3 285 751 of 15.11.1966.
19) Koch, K.H., DP 1 294 173 of 30.4.1969.
20) Hoshino, K., DAS 1 442 323 of 12.12.1968.
21) Schönwälder, H., Personal Communication.
22) Koch, K.H., DP 194 173 of 30.4.1969.
23) Ajinomoto, Brit. P. 1 190 156 of 22.1.1970.
24) Bouthillet, R.J. Food Res. 10, 201 (1951).
25) Schormüller, J., Handbuch der Lebensmittelchemie, Vol.1, P 772, Berlin-Heidelberg-New York: Springer (1965).
26) Woskow, M.H., Food Technol. 23, 1364 (1969).

27) Ney, K.H., IUPAC Symp. "The Contribution of Food Chemistry to Food Supplies", Hamburg 29.-31.8.1973, London: Butterworth, P. 411 (1974).
28) Ney, K.H., 14. CIIA Symp.: Natürliche und Synthetische Zusatzstoffe in der Nahrung des Menschen, Saarbrücken 8.-11.10.1972. Steinkopff: Darmstadt, P. 131 (1972).
29) Noguchi, M., Yamashita, M., Arai, S., Fujimaki, M., J. Fd. Sci. 40, 367 (1975).
30) Noguchi, M., Arai, S., Yamashita, M., Kato, H., Fujimaki, M., J. Agr. Fd. Chem. 23, 49 (1975).
31) Rao, C.S., Dilworth, B.C., Day, E.J., Chen, T.C., J.Fd.Sci. 40, 847 (1975).
32) Becker, E., Ney, K.H., Z. Lebensm. Unters. Forsch. 124, 206 (1965).
33) Fujimaki, M., CIIA Symp. As Lit. (28), P. 102.
34) Minaminiura, N., Matsumara, Y., Fukumoto, J., Yamamoto, T., Agr. Biol. Chem. (Tokyo) 36, 588 (1972).
35) Kauffmann, T., Kossel, Ch., Biochem. Z, 331, 377 (1959).
36) Polzhofer, K.P., Hoppe-Seyler's Z. Physiol. Chem. 352, 1 (1971).
37) Fujimaki, M., Yamashita, M., Okazawa, Y., Arai, S., Agr. Biol. Chem. 32, 794 (1968).
38) Fujimaki, M., Yamashita, M., Okazawa, Y., Arai, S., J. Fd. Sci. 35, 215 (1970).
39) Yamashita, M., Arai, S., Fujimaki, M., Agr. Biol. Chem. 33, 321 (1965).
40) Arai, S., Yamashita, M., Kato, H., Fujimaki, M., Agr. Biol. Chem. 34, 729 (1970).
41) Kirimura, J., Shimru, A., Kimizuka, A., Ninomiya, T., Katsuya, J., J. Agr. Fd. Chem. 17, 689 (1969).
42) Arai, S., Yamashita, M., Fujimaki, M., Agr. Biol. Chem. 36, 1253 (1972).
43) Solms, J., 4. Europ. Symp. "Lebensmittel-Fortschritte in der Verfahrenstechnik", Dechema-Monographien 1327/1350, Band 70, S. 337, Weinheim, Verl. Chemie 1972.
44) Thomasow, J., Deutsche Milchwirtschaft, Hildesheim 46, 2004 (1972).
45) Wieser, H., Belitz, H.D., Zeitschr. Unters. Lebensm. Forsch. 159, 329 (1975).
46) Huber, L., Klostermeyer, H., Milchwissenschaft 29, 449 (1974).
47) Guigoz, V., Solms, J., Lebensm. Wiss. u. Technol. 7, 356 (1974).
48) Sparrer, D., Petrischeck, A., Mitt. B. GDCh Fachgr. Lebensm. Chem. u. ger. Chem. 28, 279 (1974).
49) Clegg, K.M., Lim, C.L., Manson, W., Dairy Res. 41, 238 (1974).
50) Sparrer, D., Belitz, H.D., Z. Lebensm. Unters. Forsch. 157, 197 (1975).

51) Schalinatus, E., Behnke, U., Nahrung 18, 697 (1974).
52) Eriksen, S., Fagerson, I.S., Flavours 7, 13 (1976)
53) Pickenhagen, W., Forsch. Kreis Ernährungsind. 6.11.75, Hannover.
54) Guigoz, Y., Solms, J., Chemical Senses and Flavour 2, 71 (1976).
55) Beets, M.G.J., Structure-Activity Relationships in Human Chemoreception, P. 309 ff., Appl. Sci. Publ. London 1978.
56) Lalasidis, G., Sjöberg, L.B., J. Agr. Fd. Chem. 26, 742 (1978).
57) Visser, S., Slangen, K.J., Hup, G., Neth. Milk Dairy J. 29, 319 (1975).
58) Richardson, B.C., Creamer, L.K., N.Z.J. Dairy Sci. Techn. 8, 46 (1973).
59) Adler-Nissen, J., J. Agr. Fd. Chem. 24, 1090 (1976).
60) Visser, F.M.W., Neth. Milk Dairy J. 31, 265 (1977).
61) Mercier, J.-C., Grosclaude, F., Ribadeau-Dumas, B., Milchwissenschaft 27, 402 (1972).
62) Pelissier, J.-P., Mercier, J.-C., Ribadeau-Dumas, B., Ann. Biol. Anim. Bioch. Biophys. 14, 343 (1974).
63) Mercier, J.-C., Grosclaude, F., Ribadeau-Dumas, B., Eur. J. Biochem. 23, 41 (1971).
64) Gordon, W.G., Groves, M.L., J. Dairy Sci. 50, 574 (1975).
65) Matoba, T., Hayashi, R., Hata, T., Agr. Biol. Chem. (Tokyo) 34, 1235 (1970).
66) Belitz, H.D., Sparrer, D., Lebensm. Wiss. u. Technol.4, 131 (1971).
67) Hamilton, J.S., Hill, R.D., Van Leuwen, H., Agr. Biol. Chem. 38, 375 (1974).
68) Shiraishi, H., Okuda, K., Sato, Y., Yamaoka, N., Tuzimura, K., Agr. Biol. Chem. 37, 2427 (1973).
69) Fujimaki, M., Yamashita, M., Okazawa, Y., Arai, S., Agr. Biol. Chem. 32, 794 (1978).
70) Fujimaki, M., Yamashita, M., Okazawa, Y., Arai, S., J. Fd. Sci. 35, 215 (1970).
71) Wieser, H., Belitz, H.D., Z. Lebens. Unters. Forsch. 159, 329 (1975).
72) Ney, K.H., Z. Lebens. Unters. Forsch. 149, 321 (1972).
73) Petrischek, A., Lynen, F., Belitz, H.D., Lebens. Wiss. u. Technol. 5, 47 (1972).
74) Petrischek, A., Lynen, F., Belitz, H.D., Lebensm. Wiss. u. Technol. 5, 77 (1972).
75) Arai, S., Yamashita, M., Kato, H., Fujimaki, M., Agr. Biol. Chem. 34, 729 (1970).
76) Pilnik, W., Gordian 73, 208 (1973).
77) Yamashita, M., Arai, S., Matsuyama, J., Gouda, M., Kato, H., Fujimaki, M., Agr. Biol. Chem. 34, 1484 (1970).
78) Yamashita, M., Arai, S., Matsuyama, J., Kato, H., Fujimaki, M., Agr. Biol. Chem. 34, 1492 (1970).

79) Eriksen, G., Fagerson, I.S., J. Fd. Sci. 41, 490 (1976).
80) Ney, K.H., Fette-Seifen-Anstr.mittel 80, 323 (1978).
81) Wieser, H., Belitz, H.D., Z. Lebensm. Unters. Forsch. 160, 383 (1976).
82) Ney, K.H., 14th World Congr. of the Int. Soc. for Fat Research, Brighton 17.-22.10.1978. Abstract No. 0108.
83) Ney, K.H., Fette-Seifen-Anstr.mittel, under press.
84) Baur, C., Grosch, W., Wiesner, H., Jugel, H., Z.Lebensm. Unters. Forsch. 164, 171 (1977)
85) Wieske, Th., Guhr, G., Personal Communication.
86) Birch, S.G., Lindley, M.G., J. Fd. Sci. 38, 665 (1973).
87) Birch, S.G., Lee, C.K., J. Fd. Sci. 41, 1403 (1976).
88) Oberdieck,R., Riechstoffe-Aromen-Kosmetika 27, 120 (1977).
89) Oberdieck,R., Riechstoffe-Aromen-Kosmetika 27, 153 (1977).
90) Boudreau, J.C., M.B.A.A., Techn. Quarterly 15, 94 (1978).
91) Kionka, H. Strätz, F., Archiv f. Exptl. Pathol. u. Pharmakol. 95, 241 (1922).
92) Solms, J., Int. Z. Vit. Forsch.39, 320 (1969).

RECEIVED August 2, 1979.

7

Taste Components of Potatoes

J. SOLMS and R. WYLER

Department of Food Science, Swiss Federal Institute of Technology, 8092 Zurich, Switzerland

The composition of fresh potatoes is presented in Table I. The percent data listed for the different compounds are not more than average indications, as there are big variations in the composition of potatoes. - One question is immediately apparent: Is there a potato taste at all, and what are the corresponding compounds ? Or is the flavor quality of potatoes embodied in the volatile fraction ?

Indeed, potatoes are rather neutral in flavor, but they contain typical taste and odor substances. Their overall acceptance in the U.S. and in Europe is very high (1), higher than for many other commodities. A bland food would never obtain such a high acceptability. However, according to Burr (2) none of the four primary taste sensations of sour, salty, sweet and bitter is ordinarily perceptible in normal cooked potatoes.

Looking at compounds with direct taste effects, the significance of amino acids and nucleotides in the formation of potato taste has been described in several papers (3,4,5). The free amino acids and 5'-nucleotides are certainly an important fraction; they contribute to taste due to their content of glutamic acid, aspartic acid, 5'-AMP, 5'-IMP and other compounds. From the vast literature two analytical examples which have also been tested in taste tests are presented in Table II.

Buri et al. (3) presented results of taste tests with synthetic mixtures comparable to the natural systems. The important nonvolatiles were grouped as listed in Table II (nucleotides, glutamic acid, and other free amino acids), reconstituted stepwise, and tested with a ranking test. The results gave an increase in taste quality with each step with results of high significance. The final mixture had an agreeable basic taste, corresponding to the basic taste of potatoes. Moreover, the taste of the different tested potato varieties (Bintje, Ostara) were different in character. Since no other plant food is known that contains such a high amount of 5'-nucleotides, especially 5'-GMP, this point has been investigated in some detail.

Raw potatoes contain only very small amounts of 5'-nucleotides and no 5'-GMP. Since ribonucleic acid is the only possible precursor present, an en-

0-8412-0526-4/79/47-115-175$05.00/0

Table I. Composition of fresh potatoes

Compounds	Average values in %
Starch	10.0
Amylose	
Amylopectin	
Proteins	2.0
Organic acids	1.5
Citric acid	
Malic acid	
Succinic acid	
Fumaric acid	
Minerals	1.0
K, Mg, Ca, P, Na	
Amino acids (free)	0.8
(all "current" amino acids)	
Non-starch Polysaccharides	0.7
Hemicelluloses	
Pectins	
Hexosans	
Pentosans	
Sugars	0.5
Glucose	
Fructose	
Saccharose	
Lipids	0.2
(diverse fractions)	
Polyphenols	0.2
Chlorogenic acid	
Caffeic acid	
Vitamins	0.02
Ascorbic acid etc.	
Pigments	0.015
Anthocyans	
Carotinoids	
Alcaloids	0.01
Solanins	
Chaconins	
RNA, Nucleotides	0.01

Table II. Free amino acids and nucleotides of boiled potatoes of the varieties Bintje and Ostara (in parenthesis) in mg per 100g fresh material. From Solms (4).

Amino acids:

I. Glu 73.8 (36.4)

II. Ala 10.1 (7.4), Arg 19.8 (19.0), Asp-NH_2 220.0 (187.0), Asp 46.8 (36.4), Cys-S- 1.2 (0.5), Glu-NH_2 49.2 (77.6), Gly 2.2 (2,4), His 4.2 (4.3), i-Leu 10.6 (6.0), Leu 6.1 (2.9), Lys 6.8 (5.6), Met 9.2 (6.3), Phe 11.8 (4.5), Pro 9.1 (5.2), Ser 6.4 (6.4), Thr 8.0 (8.0), Try 3.0 (0.9), Tyr 11.0 (4.8) Val 25.8 (16.8)

Total amino acids: 535.1 (438.4)

Nucleotides:

III. 5'-AMP 3.0 (2.25), 5'-GMP 2.11 (1.39), 2'3'-GMP 1.72 (1.79), 5'-UMP 2.14 (1.78)

Total nucleotides: 8.96 (7.22)

zymatic hydrolysis of RNA during heat processing is the most likely source of the nucleotides (6) (Table III). It is well known that RNA degrading enzymes are present in potato tissue as well as in any other plant material. However, typical pH and temperature conditions in the potato tubers during heating seem to be responsible for the liberation of the 5'-nucleotides in sufficient amounts. It is known that the temperature of potato tissue rises during heating very slowly through the 40-60°C region (7). - The enzymes investigated in connection with this process (they are all responsible for RNA attack in the tuber) are summarized in Table IV. A combination of their activities during heat processing can be summarized as follows.

RNA degradation up to 50°C gives a preferential activity of phosphodiesterase I, and leads to the liberation of 5'-nucleotides. An increased activity of enzymes, liberating 2'3'-nucleotides, can be found only at higher temperatures and a later stage, when the substrate has already been used up by phosphodiesterase I.

The nucleotide attacking enzymes, which would destroy the formed 5'-nucleotides, are similar in behaviour with respect to temperature but differ in pH optimum of the tuber, and therefore remain rather inactive.

The result is an optimum accumulation of 5'-nucleotides around 50°C and pH 6.0 occuring in a temperature gradient during heating and leading to a final inactivation of all enzymes.

An experiment with a potato enzyme raw extract and RNA as substrate is presented in Table V. Autoincubation experiments with potato tissue confirm these results and are shown in Table VI (8).

It can be summarized that the accumulation of 5'-nucleotides is due to the natural enzymes present which are selectively active under specific pH and temperature conditions during the heating of the tubers. It is certainly of interest to take these results into consideration for the industrial utilization of potatoes and the acceptability of the product.

The sugars glucose, fructose and saccharose occur in varying concentrations, depending on the physiological state of the potato tuber. They are important compounds participating in desired or undesired browning reactions but have apparently no positive contribution to the potato taste. If they occur in relatively large amounts and confer a sweet taste to the product, the acceptability of the food is reduced. The sweet taste quality seems to be not a desirable one in potato taste systems (9,10). Sinden, Deahl and Aulenbach discuss the earlier literature and investigate in their own experiments the importance of bitterness and astringency, the most frequently noted off-flavors in potatoes (11,12,13). They found a correlation between bitterness and astringency and glycoalkaloid content, but no relation to the polyphenol content. Nothing detailed is known about the taste contribution of the minerals fraction, although it is probable that it contributes some effect. The potato proteins are tasteless, however, they are rich in hydrophobic amino acids, and therefore can form bitter tasting peptides on hydrolysis (14,15). There is, how-

Table III. Free and bound nucleotides in raw and boiled potatoes (μMol/kg potatoes). From Buri & Solms (6).

Free nucleotides	Raw potatoes 2', 3'	5'	Boiled potatoes 2', 3'	5'
Uridinemonophosphate	26.9[a]		39.2	68.2
Adenosinemonophosphate	0	5.5	0	110.3
Guanosinemonophosphate	0	0	26.7	64.4
Cytidinemonophosphate	0	0	0	26.5
Adenosinediphosphate		26.2		37.2
Adenosinetriphosphate		22.8		25.9
Bound nucleotides (presumably RNA)	577.6		133.6	

[a] Present as sugar nucleotides

Naturwiss

Table IV. RNA degrading enzymes in raw extract from potatoes From Dumelin & Solms (8).

	Optimum activities °C	pH
Phosphodiesterase I	50	5.5
Phosphodiesterase II	60	5.5
Ribonuclease	70	5.0
Phosphatase	45 - 50	5.0
5'-Nucleotidase (+ Phosphatase)	45 - 50	5.0

Table V. Release of soluble degradation products from yeast RNA during incubation with raw enzyme extract from potatoes at different temperatures and pH values (degradation products in µM x 10^{-3}/ml). From Dumelin & Solms (8).

Temperature (°C)												
pH	5.0	5.5	6.0	6.5	5.0	5.5	6.0	6.5	5.0	5.5	6.0	6.5
total nucleotides	39	55	110	146	66	66	203	154	164	120	136	104
total nucleosides	78	177	141	94	222	244	192	96	222	210	143	38
5'-nucleotides	20	36	74	55	41	52	130	69	23	34	110	48
3'-nucleotides	19	19	36	91	25	14	73	85	141	86	26	56

Potato Research

Table VI. Formation of 5'-nucleotides and 3'-nucleotides during autoincubation of potato tissue homogenate at 52°C. From Dumelin & Solms (8).

	Nucleotides in µM/kg fresh tissue		
pH	5.0	6.0	6.5
5'-nucleotides	485	1138	628
3'-nucleotides	119	12	243
total nucleotides	604	1150	871

Potato Research

ever, no evidence that this occurs normally in potato products for food purposes.

The largest fraction occuring in potatoes is the starch fraction. Starch forms the matrix of all potato products. Starch has probably no taste of its own but it has indirectly a great influence on flavor with tactile and other effects. Considerable work has been reported on the gelatinization of starch and the textural characteristics of gelatinized starch (16). Instrumental methods for the measurement of texture of potato products, especially mashed potato, with rheological parameters and possible relationships with sensory data have been reported only recently (17,18). It is also known that gelatinized starch can form inclusion complexes under helix formation with various compounds (19, 20). In our experiments potato starch seemed to be a most effective compound in forming inclusion complexes (21).

Generally the inclusion reaction is described to take place in a thermal gradient ranging from 90°C to room temperature. The complexes formed are often insoluble and can be separated as precipitates (21, 22). Inclusion complexes such as these often form under normal food processing conditions. The complexing of free starch due to the addition of fatty acid derivatives during production to potato flakes for instant mashed potatoes is a case in point. In this case the desired effect is related to taste due to a perceptible change in texture.

Recent experiments in our laboratory have shown that a thermal gradient is not necessary for complex formation with potato starch. In the presence of suitable compounds, the gelatinized potato starch forms helices under complex formation at isothermal conditions. The reaction takes place even at low concentrations of ligand compounds under conditions occuring in any food system containing potato starch. The reaction can easily be followed by amperometric titration with iodine (24). The complexes formed can be analyzed by using a combination of glucose determination and G.C. analysis.

An example for such a reaction with decanal as ligand and potato starch as complexing compound is discussed in the following:

The formation of decanal - starch - complexes with time at different temperatures is presented in Figure 1. The reaction is completed in a matter of minutes, and a stable equilibrium is obtained. Isotherms of complex formation are shown in Figure 2. The complex formation starts at very low concentrations and comprises appreciable amounts of ligand. We are actually studying these reactions in some detail and are interested in the consequences of these interactions for taste and odor perception. As reported elsewhere, the complexed ligands, if present in dry state, have a remarkably increased chemical stability (25).

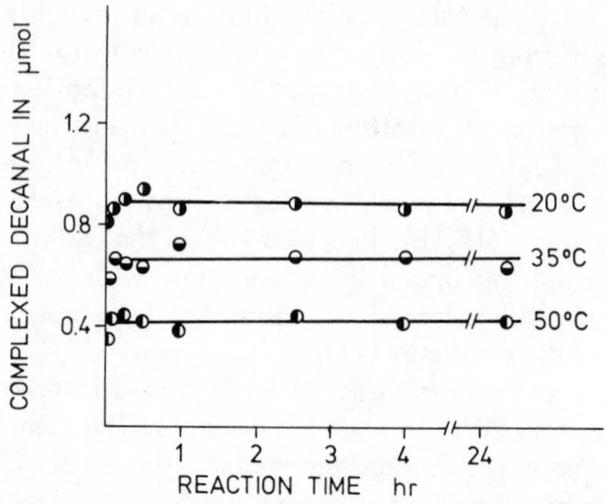

Figure 1. Formation of decanal–starch complexes with time—starch (as glucose): 0.613 mM; decanal: 5.31 μM; H_2O: 10 mL; pH: 7.0; temperature: 20, 35, 50°C

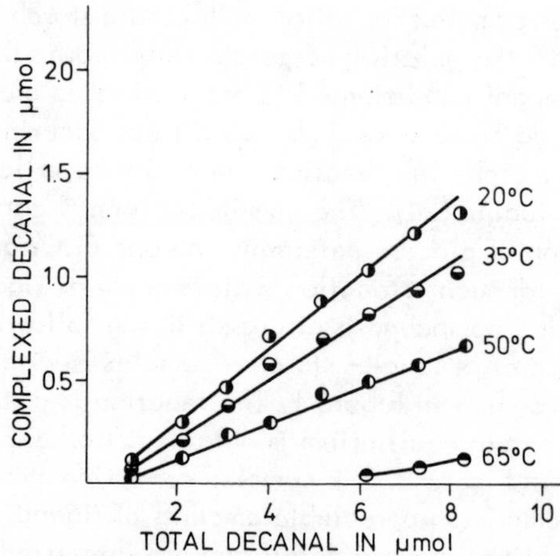

Figure 2. Isotherms of complex formation of potato starch with decanal—starch (as glucose): 0.613 mM; decanal: variable; H_2O: 10 mL; pH: 7.0; temperature: 20–65°C

Conclusions and summary

Potato taste is not characterized by one of the primary taste sensations. Especially sweet, sour or bitter notes are considered off-flavors. However, free amino acids and 5'-nucleotides are important compounds that convey an agreeable basic taste to potato products. The amino acids occur naturally in free form; the 5'-nucleotides are liberated during the heat preparation of potatoes by a specific enzymatic degradation of RNA. Starch forms a matrix for all potato preparations. Although it is tasteless, is has an influence on taste quality due to textural characteristics, and due to its pronounced capability to form stable complexes with flavor compounds either in a thermal gradient or under isothermal conditions.

Literature Cited

1. Harper, R., Nature, 1963, 200, 14.
2. Burr, H.K., in "Proc. Plant Science Symposium", Campbell Institute for Agric. Research, Camden, NJ, 1966, p. 83.
3. Buri, R., Signer, V. and Solms, J., Lebensm. Wiss. Technol., 1970, 3,63.
4. Solms, J., in Ohloff, G.F., Thomas, A.F., "Gustation and Olfaction", Academic Press, London, New York, 1971, p. 92.
5. Lipsits, D.V. and Sikilinde, V.A., Appl. Biochem. Microbiol. (USSR), 1972, 8, 225.
6. Buri, R. and Solms, J., Naturwiss., 1971, 58, 56.
7. Buri, R. "Ueber das Vorkommen von Nukleotiden in Kartoffeln und ihre Bedeutung für den Flavor", Thesis, Swiss Federal Institute of Technology, Zurich, No. 4647, 1971.
8. Dumelin, E. and Solms, J., Potato Res., 1976, 19, 215.
9. Kröner, W. and Völksen, W., "Die Kartoffel", J.A. Barth Verlag, Leipzig, 1950, p.95.
10. Burton, W.G., "The potato", H. Veenman & Zonen, Wageningen, Holland, 1966, p. 183.
11. Mondy, N.I., Metcalf, C. and Plaisted, R.L., J. Food Sci., 1971, 36,459.
12. Sinden, S.L., Deahl, K.L. and Aulenbach, B.B., J. Food Sci., 1976, 41, 520.
13. Mondy, N.I., Metcalf, C., Hervey, J. and Plaisted, R.L., in "Proc. 19th Nat. Potato Utiliz. Conf.", ARS-USDA, 1969, 73.
14. Ney, K.H., Fette Seifen Anstrichmittel, 1978, 80, 323.
15. Ney, K.H., in "Proc. World Conf. on Vegetable Proteins", Amsterdam, 1978.
16. Whistler, R.L. and Paschall, E.F., "Starch, Chemistry and Technology" I & II, Academic Press, London, New York, 1967.
17. Schweingruber, P., Escher, F. and Solms, J., in Sherman, P., "Food Texture and Rheology", IUFoST Symposium, Academic Press, New York, 1979, in press.

18. Schweingruber, P., Escher, F. and Solms, J., Mitt. Gebiete Lebensm.Hyg., 1979, in press.
19. Foster, J.F., in Whistler, R.L. and Paschall, E.F., "Starch, Chemistry and Technology" I, Academic Press, London, New York, 1967.
20. Senti, F.R., and Erlander, S.R., in Mandelcorn, L., "Non-Stoichoimetric Compounds", Academic Press, London, New York, 1964, p. 588.
21. Osman-Ismail, F., "The formation of inclusion compounds of starches and starch fractions", Thesis, Swiss Federal Institute of Technology, Zurich, No. 4829, 1972.
22. Osman-Ismail, F. and Solms, J., Stärke, 1972, 24, 213.
23. Talburt, W.F., and Smith, O., "Potato Processing", AVI Publ. Company, Westport, Conn., 1967.
24. Hollo, J. and Szejtli, J., Stärke, 1956, 8, 123.
25. King, B., Wyler, R. and Solms, J., in Land, D.G. and Nursten, H.E., "Progress in Flavour Research, 2nd Weurman Flavor Symposium, Proceedings", Applied Science Publishers Ltd., Barking GB, 1979.

Acknowledgements

This work was supported, in part, with funds from the Eidgenössische Alkoholverwaltung, Berne.

RECEIVED August 7, 1979.

8

The Taste of Fish and Shellfish

SHOJI KONOSU

Laboratory of Marine Biochemistry, Faculty of Agriculture,
The University of Tokyo, Tokyo, Japan

In Japan, a wide variety of marine products, such as algae, molluscs, crustaceans, echinoderms, and fish have been consumed with relish from olden times. These food habits have stimulated many studies on the extractive components which may contribute to the taste of these products. Several comprehensive reviews on the subject are available (1-8). In order to avoid overlapping with them, special references are made in this review to those components whose roles in producing the taste of fish and shellfish have been examined organoleptically.

Taste-active Components in Fish

Enormous efforts have been devoted to the analysis of the extractive components of fish muscles and much information has been accumulated. In recent years, the distribution of nitrogenous components in the muscle extracts of several species of fish has been elucidated almost completely (9, 10, 11, 12, 13). However, few studies have correlated these analytical data directly with taste.

In this section, only the taste-producing properties of hypoxanthine and histidine in fish will be reviewed. For other components, refer to the excellent reviews by Jones (14, 15).

Hypoxanthine. In fish muscles, IMP is accumulated as a post mortem degradation product of muscle ATP. It has been postulated by Hashimoto (2) that IMP thus accumulated, in combination with glutamic acid, forms the nucleus of the taste of fish meat. IMP is then slowly degraded to hypoxanthine through inosine. According to Jones (14), inosine was barely detectable by trained or untrained palates at the maximum concentrations present in cod muscle, and description of the taste ranged from sweet to acid-astringent. Unlike inosine, however, hypoxanthine has a strongly bitter taste. Jones (14) has described the bitter taste of cod muscle appearing after chill storage for 10 days as being attributable to hypoxanthine.

0-8412-0526-4/79/47-115-185$05.00/0

In this connection, an interesting property of hypoxanthine has been found by Spinelli (16). In dilute solutions it produced a variety of taste sensations, the predominating ones being bitterness or dryness. Eight of ten panelists found it to be bitter at a concentration of 0.01% in distilled water. Addition of 3-6 μmol of hypoxanthine to one g (0.4-0.8%) of fresh or irradiated and stored petrale sole, *Eopsetta jordani*, having a bacterial count of less than 10^6 / g, did not produce a consistently detectable change in flavor. However, when the bacterial counts exceeded 10^6/g, a change was detectable. From these results, Spinelli suspected that bacterial growth changed the flavor characteristics of hypoxanthine by utilizing or altering some constituents of fish muscle that normally mask its flavor, or by producing metabolic products that enhance its flavor. This finding cautions us that the taste potency of a component in foods should not be assessed solely from its taste in pure solution.

Histidine. As shown in Table I, scombroid fish, such as tuna, skipjack, and mackerel, contain a large amount of free histidine in their muscle (7). Opinions on the contribution of

Table I. Free amino acids in the muscle of some scombroid fish (mg/100g)

	Mackerel	Big-eye tuna	Yellowfin tuna	Skipjack	Yellowtail
Glycine	15.8	11.0	3.1	8.9	3.7~ 6.1
Alanine	22.2	21.5	6.6	22.6	13.9~27.5
Valine	1.4	14.3	6.7	4.1	2.6~10.2
Leucine	4.7	10.8	7.1	3.4	3.1~12.4
Isoleucine	0.9	5.8	3.1	2.0	1.8~ 6.7
Proline	-	2.0	1.6	+	0.9~48.2
Phenylalanine	3.0	4.6	1.5	2.5	1.9~ 4.7
Tyrosine	5.5	5.5	2.0	2.5	1.7~ 6.1
Serine	+	5.2	2.0	3.1	4.3~ 6.8
Threonine	8.1	7.7	3.0	3.8	2.9~10.9
Methinonine	2.5	9.0	3.1	1.4	- ~ +
Arginine	+	0.4	0.6	-	
Histidine	781	745	1220	1110	1010~1220
Lysine	17.1	3.8	35.2	11.2	61.7~90.1
Aspartic acid	2.3	1.0	1.1	2.9	
Glutamic acid	17.8	19.9	3.3	7.0	5.1~27.9
Taurine	+	21.1	26.4	16.1	25.1~89.7

+, trace; -, not detected.
From Suyama (7).

Suisangaku Series

this amino acid to the flavor, however, vary. Simidu *et al*. (17, 18) postulated that the amino acid may participate in the flavor of these fish, since the more palatable species contain more free histidine in the muscle, and the post mortem changes in the palatability of such fish as tuna and skipjack run parallel to the changes in their free histidine content. Endo et al. (19) have also reported that the difference in palatability between aqueous extracts from the muscles of cultured and wild yellowtails may be attributable to the difference in their free histidine content, because, of the extractive components analyzed, only in histidine was there a significant difference between cultured and wild fish.

On the other hand, Hughes (20) has stated that the addition of 400 mg of histidine to 100 g of herring meat did not produce any detectable change of flavor, when tasted after heating. The author and coworkers (21) have found in the omission test of a synthetic extract (Table II) simulating the extract of dried skipjack (katsuwobushi) that histidine, which is the most abundant amino acid, making up about 80% of the total free amino acids, did not contribute appreciably to the taste. They have also confirmed that histidine inosinate, to which umami (monosodium L-glutamate-like taste) of katsuwobushi has been attributed by Kodama (22), was indistinguishable from disodium inosinate in taste potency.

Table II. Composition of a synthetic extract simulating katsuwobushi extract (mg/100 ml)*

Histidine	90.9	Serine	0.5
Taurine	10.8	Threonine	0.5
Lysine	2.6	Isoleucine	0.4
Alanine	1.6	Aspartic acid	0.3
Glutamic acid	1.3	Methionine	0.3
Leucine	0.9	Tyrosine	0.1
Glycine	0.8	5'-IMP	19.0
Phenylalanine	0.6		
Valine	0.6	pH	5.8

* The concentration is equivalent to that of an aqueous extract prepared using 3% of katsuwobushi.
From Konosu et al. (21).

Bulletin of the Japanese Society of Scientific Fisheries

Iimura and Umeda (23) have described the free histidine occurring in quantity in katsuwobushi as serving as a taste enhancer in conjunction with lactic acid and KH_2PO_4 by elevating buffering capacity.

The role of histidine in making up the taste of scombroid fish seems to be a subject in need of further study.

Organic Acids in Shellfish

Since the pioneering work of Aoki (24) on the flavor components of shellfish, succinic acid had been believed to be the key substance responsible for their palatable taste. However, some questions have arisen about it.

Firstly, Takagi and Simidu (25) examined the correlation between the organic acid content of 9 species of shellfish and their taste, and found that the more palatable species were not necessarily richer in succinic acid, as exemplified by the hard clam, Meretrix lusoria. These results led them to conclude that succinic acid does not dominate the delicious taste of shellfish. Secondly, Konosu et al. (26) reported that the succinic acid content of the short-necked clam, Tapes japonica, when determined immediately after collection, was very low (20-40 mg/100 g of edible part) as compared with Aoki's value (330 mg), and that the flavor of a fresh sample was as good as that of a commercially available sample that had accumulated a large amount of succinic acid.

On the other hand, Take and Otsuka (27) stated that the aqueous extract of the corbicula, Corbicula leana, was judged more acceptable by testers than that from which the organic acids had been removed by extraction with diethyl ether. They reported that exceedingly large amounts of citric, malic, and glycolic acids and a small amount of succinic acid were contained in their sample. Therefore, the contribution of succinic acid to the taste is obscure. Take and Otsuka noted that a synthetic mixture (Table III) containing amino acids and organic acids in the same relative concentrations as they occurred in the corbicula extract, simulated the taste of the natural extract.

Table III. Composition of a synthetic extract simulating corbicula extract (mg/100 ml)*

Glutamic acid	32.80	Arginine	0.75
Glycine	3.33	Histidine	0.58
Isoleucine	2.58	Aspartic acid	0.57
Leucine	1.86	Succinic acid	19
Valine	1.56	Citric acid	1570
Phenylalanine	0.88	Malic acid	1000
Lysine	0.86	pH	7.3

* The concentration is equivalent to that of an aqueous extract prepared using 10% of the soft part of corbicula.

From Take and Otsuka (27)

Memoirs of the Faculty of Education, Niigata University

Thus, the role of succinic acid in making up the taste of bivalves remains to be elucidated.

Taste-active Components in Abalone

Abalone meat is highly esteemed in the Far Eastern countries. The author and Hashimoto (28) organoleptically surveyed taste-active components of abalone meat, *Haliotis gigantea discus*, by the omission test using a synthetic extract which was formulated on the basis of the analysis of the meat extract (Table IV).

Table IV. Composition of a synthetic extract simulating abalone extract (mg/100 ml)*

Taurine	946	Tyrosine	57	5'-AMP	90
Arginine	299	Valine	37	5'-ADP	12
Glycine	174	Phenylalanine	26	Glycine betaine	975
Glutamic acid	109	Leucine	24		
Alanine	98	Histidine	23	Trimethylamine oxide	3.2
Serine	95	Tryptophan	20		
Proline	83	Isoleucine	18	Trimethylamine	1.1
Threonine	82	Methionine	13	NH_3	8
Lysine	76	Aspartic acid	9	Glycogen	7400
				pH	5.8

* The concentration is equivalent to that found in abalone meat.

From Konosu and Maeda (29).

Bulletin of the Japanese Society of Scientific Fisheries

Results obtained are summarized as follows.

1) Taurine and arginine, which account for an important part of the free amino acids, made little contribution to the taste.

2) When glycine was omitted from the synthetic extract, sweetness and *umami* decreased to some extent and the overall taste became weak, but the characteristic taste of abalone meat was still retained.

3) The effect of glycine betaine was almost the same as that of glycine.

4) When glutamic acid was removed from the synthetic extract, *umami* decreased markedly, and the characteristic taste disappeared.

5) AMP was found to contribute to *umami*.

6) Elimination of each of the other components was hardly detectable by panelists, but omission of them in a group produced a considerably weaker taste.

7) Glycogen, which is contained in abalone meat in a high concentration, showed a body effect on the taste, although glycogen itself was tasteless.

These results suggest that the taste characteristic of abalone meat is constituted basically of *umami* produced by glutamic acid and AMP, and of sweetness produced by glycine and glycine betaine. The tastes produced by these substances are harmonized, smoothed, and enhanced by glycogen. It seems curious that AMP, which is almost tasteless, contributes to *umami*, but this may be explainable by the enhancing effect of AMP on MSG (monosodium L-glutamate) observed by Toi *et al.* (30). In the muscle of marine invertebrates, AMP, instead of IMP as in fish, is accumulated as a post mortem degradation product of ATP, since the muscle lacks AMP aminohydrolase, or, if present, its activity is very low. Umami of many other marine invertebrates may well be explained by the interaction of AMP and glutamic acid.

Taste-active Components in Squids

Squids are popular sea foods in Japan. They are not only consumed raw, boiled or broiled, but are also processed to sun-dried, smoked, and fermented products.

Endo *et al.* (32) analyzed the free amino acids, trimethylamine oxide (TMAO), and glycine betaine in the mantle muscle of six species of squids (Table V), and divided the squids into three groups by the composition: *Loligo chinensis*, *L. kensaki*, and *Sepioteusthis lessoniana*, which are very rich in free amino acids, especially in glycine, were affiliated with Group 1; *Sepia esculenta*, which is moderately rich in free amino acids, with Group 2; and *Thysanoteuthis rhombus* and *Ommastrephes sloani pacificus*, which are scanty in free amino acids, but abundant in TMAO, with Group 3. As the members of Groups 1 and 2 contain considerable amounts of such sweet-tasting amino acids as glycine, alanine, and proline and have a better taste than members of Group 3, Endo *et al.* assumed that these amino acids are responsible for the palatability of squids. Furthermore, they pointed out that, although glycine betaine and TMAO had been thought to make some contribution to the taste of squids, the difference in palatability among the species is not explainable by these components, because the glycine betaine content was relatively uniform among the species and TMAO content was apparently higher in Group 3 than in Groups 1 and 2. It is desirable to confirm their postulations by organoleptic tests and to examine the contribution of nucleotides and organic acids, which they have not analyzed, to the taste.

Free Amino Acids in Prawns and Lobsters

The muscle extracts of prawns and lobsters, like many other marine invertebrates, are characterized by the presence of large amounts of free glycine. Hujita *et al.* (33, 34, 35) observed that the amount of free glycine in the muscle of these crustaceans paralleled their palatability, and suggested that this amino acid should make an important contribution to the taste. Moreover, they suspected that alanine, proline, and serine, which all have a sweet taste, may also contribute to the taste to some extent, since the correlation between the palatability of the muscles and the sum of glycine and the three amino acids is highly significant, as shown in Figure 1.

Hujita (34) noticed that the decrease in palatability of prawn muscle, *Penaeus japonicus*, which proceeded along with the lowering of freshness, was accompanied by a decrease in free glycine content. Hujita *et al.* (36) also found that the free glycine content of prawn muscle is higher in winter, when the

Table V. Nitrogenous compounds in the muscle extract of squids (N mg/100 g)

	Loligo chinensis	*Loligo kensaki*	*Sepioteuthis lessoniana*	*Sepia esculenta*	*Thysanoteuthis rhombus*	*Ommastrephes sloani pacificus*
Taurine	27.8	22.5	17.9	53.8	26.7	10.8
Hydroxyproline	2.3	-	-	5.2	-	-
Aspartic acid	0.4	-	-	-	-	-
Threonine	3.6	3.0	1.0	6.7	0.3	2.4
Serine	3.6	3.5	3.6	17.8	0.7	2.9
Glutamic acid	1.4	3.3	0.3	3.2	1.0	4.0
Proline	117.1	40.1	91.1	72.7	-	22.9
Glycine	144.4	154.5	155.1	11.8	1.8	4.5
Alanine	75.6	41.0	28.5	23.5	9.3	10.7
Cystine	-	0.2	0.4	1.7	-	2.1
Valine	1.4	1.8	0.4	2.3	0.3	2.0
Methionine	2.1	0.1	0.7	2.6	0.3	1.7
Isoleucine	0.8	1.7	0.6	1.0	0.7	1.2
Leucine	1.6	0.6	1.3	1.2	0.9	2.5
Tyrosine	0.6	1.3	0.6	0.1	-	1.1
Phenylalanine	1.0	0.7	0.2	1.0	-	0.1
Tryptophan	0.4	-	0.7	-	-	-
Histidine	8.1	1.3	4.3	3.0	2.2	16.1
Lysine	4.4	6.7	2.8	4.2	1.7	4.0
Arginine	72.3	226.2	79.0	85.5	183.8	51.4
Trimethylamine oxide	129	112	92	54	257	239
Glycine betaine	102	92	102	105	111	74

-, not detected.
From Endo *et al*. (32).

Bulletin of the Japanese Society of Scientific Fisheries

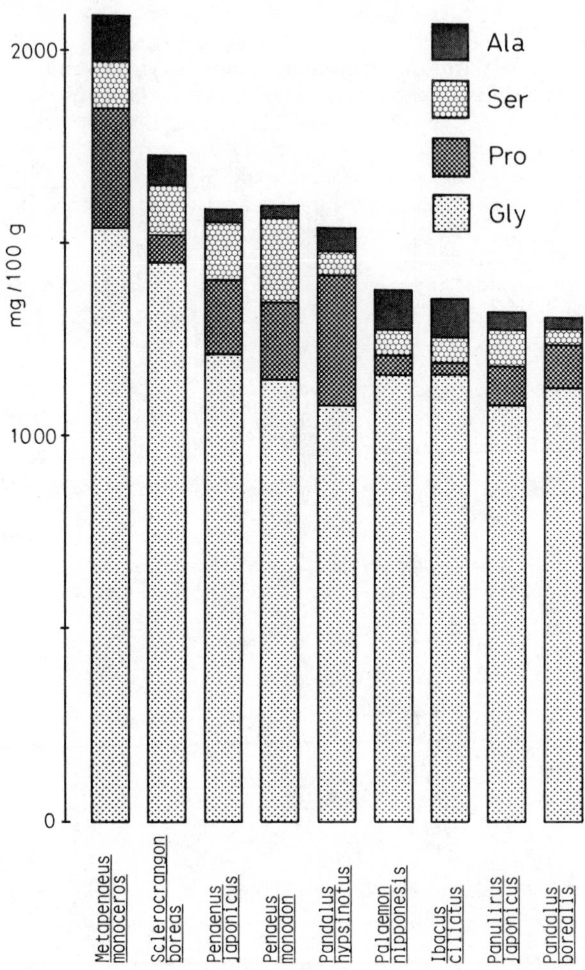

Masao Hujita (thesis)

Figure 1. The contents of glycine, proline, serine, and alanine in the muscle extracts of prawns and lobsters. They are arranged in decreasing order of palatability from left to right (34).

muscle is more palatable. These observations serve as additional evidence to support their assumption that glycine is an important taste-giving constituent in prawns and lobsters.

Using aqueous extracts of a dried shrimp, Sergestes lucens and a fresh prawn, Pandalus borealis, Take et al. (37) examined organoleptically the role of various constituents in the make-up of their taste. The results are summarized as follows.

1) When nucleotides, consisting chiefly of AMP, were degraded with nucleotidase, or when organic acids were removed with diethyl ether, no detectable change of taste occurred.

2) When the extracts were incubated with glutamate decarboxylase, their umami slightly decreased.

3) When extracts were passed through a column of Amberlite IR-120, they lost their umami completely and produced only a slightly sweet sensation.

From these results and from the free amino acid compositions of the extracts, they regarded amino acids, chiefly glutamic acid and glycine, as the main contributors to the taste of these crustaceans.

Taste-active Components in Crabs

Take et al. (38) who surveyed the taste-active components of the snow crab, Chionoecetes opilio, in the same way as described for the shrimp and prawn in the preceding section, found that the amino acids were the most important flavor components, and that a synthetic extract (Table VI) prepared by simulating the crab extract reproduced its taste fairly well.

Table VI. Composition of a synthetic extract simulating snow crab extract (mg/100 ml)*

Arginine	71.00	Tyrosine	5.54
Glycine	63.00	Glutamic acid	5.50
Valine	25.55	Aspartic acid	5.42
Alanine	21.08	Histidine	1.64
Proline	15.07	Maltose	300
Lysine	6.62		

* The concentration is equivalent to that of an aqueous extract prepared using 10% of the muscle. Maltose equivalent to the amount of reducing sugars found was added. The pH value of the synthetic extract is unknown.
From Take et al. (38).

Journal of Home Economics (Japan)

We have recently undertaken a similar, but more extensive study, to elucidate the contribution of each of the extractive components to the characteristic taste of boiled crabs. The outline of our study follows.

Five species of common edible crabs were used. They were cooked in boiling water containing 3% NaCl for 20 minutes according to commercial practice. After cooling, the leg meat was removed from the crabs of both sexes and extracted with hot water. The extracts were then deproteinized with 80% ethanol and analyzed for free and combined amino acids, nucleotides and related compounds, quaternary ammonium bases, sugars, organic acids, and inorganic ions.

As shown in Figure 2, a common striking feature of the free amino acid composition of the crab meats was the considerably high amounts of glycine and arginine and the high but somewhat lower amounts of proline and taurine. These four amino acids accounted for 60-80% of the total free amino acids, which amounted to 2,000-3,000 mg/100 g of meat. After acid hydrolysis of the extracts, the total amino acids showed a 10-20% increase, glutamic and aspartic acids accounting for the major part of the increase (39).

The amounts of the other constituents are summarized in Figures 3, 4, and 5. AMP and CMP comprised the major part of the total nucleotides. It is worthy to note that crab meat is very rich in quaternary ammonium bases, such as glycine betaine and TMAO. Appreciable amounts of homarine were also found (40). Although several kinds of sugars and organic acids were detected, their concentrations were extremely low, except for lactic acid, which ranged from 30 to 200 mg/100 g and glucose, which ranged from 3 to 86 mg/100 g. In the case of the inorganic ions, sodium, potassium, chloride, and phosphate ions were predominant.

All of these analytical results are summarized in Figure 6, in which the amount of each extractive component is shown in terms of the percentage of the total extractive nitrogen and in terms of the percentage of the dry matter of the extracts. In both cases, recoveries were satisfactory, reaching more than 92%. These values indicate that the composition of the crab meat extracts has been elucidated almost completely. From these results a synthetic extract (Table VII) simulating the composition of the meat extract of the male snow crab, which is reputed to be one of the most palatable species in Japan, was prepared. After confirming organoleptically that the synthetic extract could reproduce the taste of the crab extract fairly well, omission tests were carried out in order to determine how each constituent contributed to the taste. Using the triangle difference test a team of seven trained panelists compared the taste of the synthetic extract lacking certain ingredient(s) to the taste of the complete synthetic extract.

The results showed that seven nitrogenous constituents, glycine, alanine, glutamic acid, arginine, AMP, GMP, and CMP, and four inorganic ions, Na^+, K^+, Cl^-, and PO_4^{3-}, contribute more or less to produce the taste of crab. The opinions of the panelists regarding the taste of the test solutions are summarized as follows.

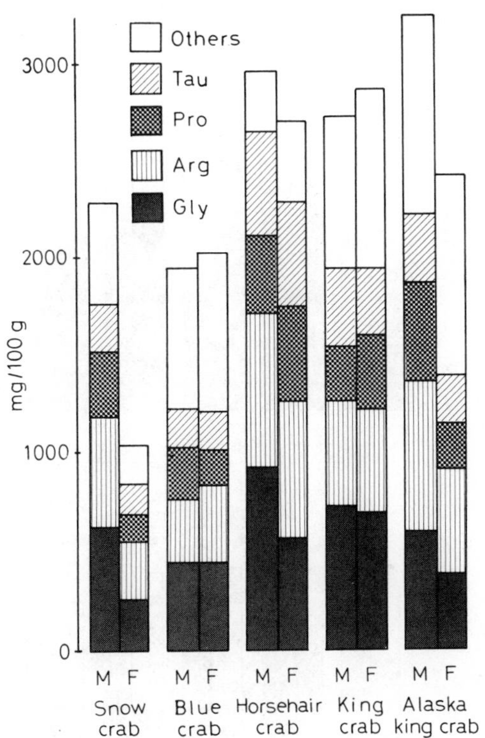

Figure 2. Free amino acids in crab extracts

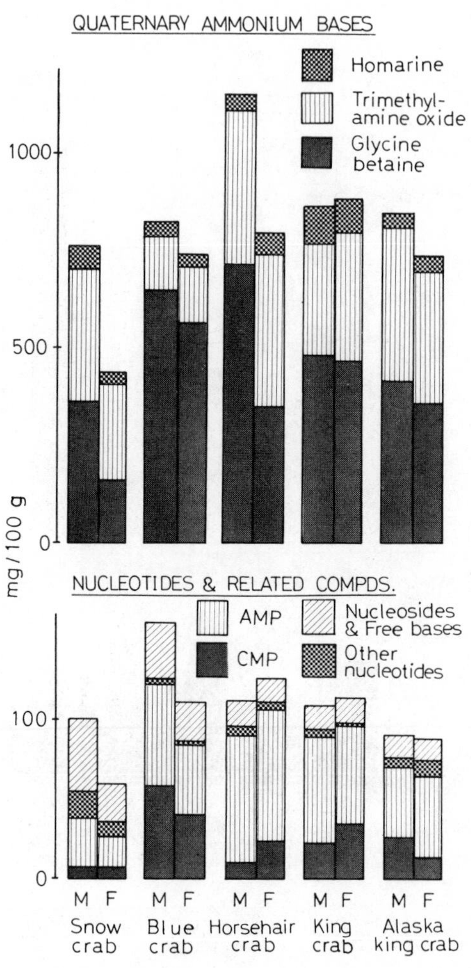

Figure 3. Nitrogenous components in crab extracts

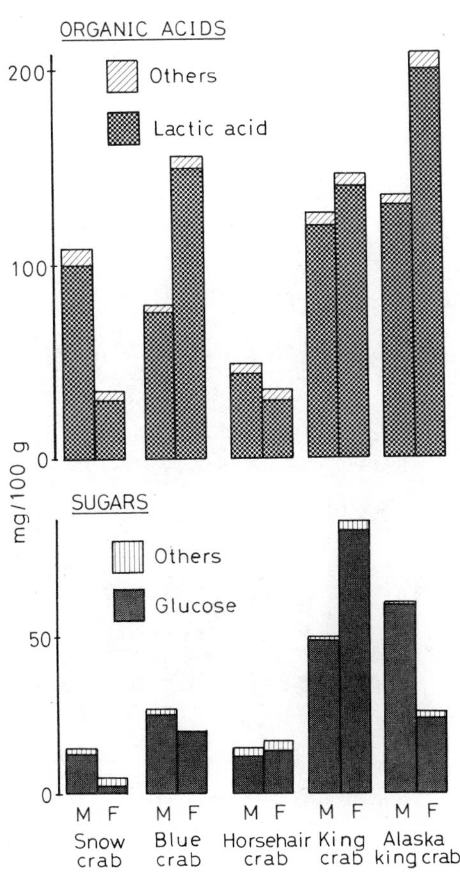

Figure 4. Sugars and organic acids in crab extracts

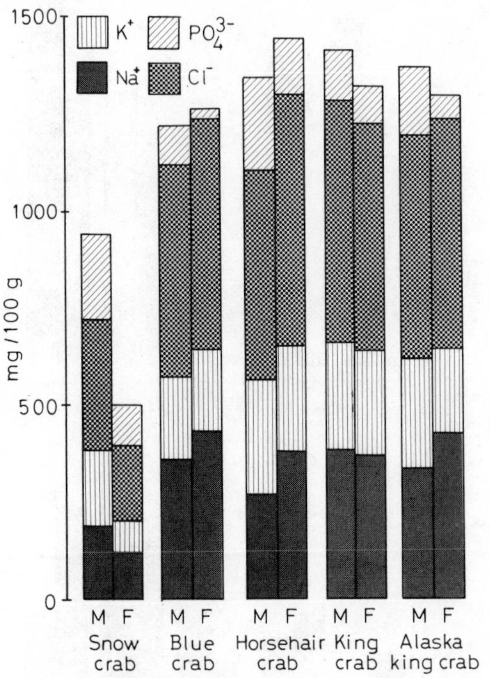

Figure 5. Inorganic components in crab extracts

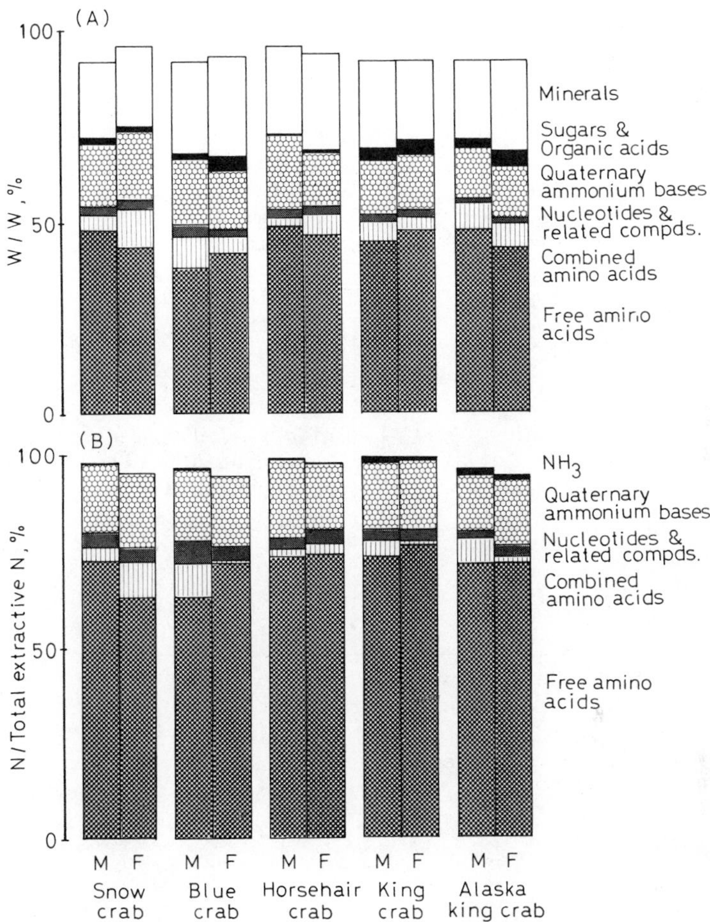

Bulletin of the Japanese Society of Scientific Fisheries

Figure 6. Distribution of various components in crab extracts: (A) weight basis; (B) nitrogen basis (40)

Table VII. Composition of a synthetic extract simulating snow crab extract (mg/100 ml)*

Taurine	243	5'-CMP	6
Aspartic acid	10	5'-AMP	32
Threonine	14	5'-GMP	4
Serine	14	5'-IMP	5
Sarcosine	77	5'-ADP	7
Proline	327	Adenine	1
Glutamic acid	19	Adenosine	26
Glycine	623	Hypoxanthine	7
Alanine	187	Inosine	13
α-Amino-n-butyric acid	2	Guanine	1
		Cytosine	1
Valine	30	Glycine betaine	357
Methionine	19		
Isoleucine	29	Trimethyl-amine oxide	338
Leucine	30		
Tyrosine	19	Homarine	63
Phenylalanine	17	Glucose	17
Ornithine	1	Ribose	4
Lysine	25	Lactic acid	100
Histidine	8	Succinic acid	9
3-Methyl-histidine	3	NaCl	259
		KCl	376
Tryptophan	10	NaH_2PO_4	108
Arginine	579	Na_2HPO_4	569
		pH	6.60

* The concentration is equivalent to that found in the crab meat. For organoleptic tests, the solution was diluted twice.

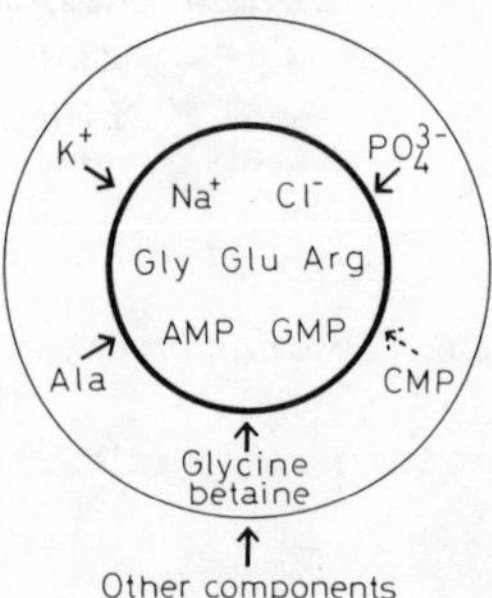

Figure 7. Model for the construction of crab taste

1) When glycine is omitted from the synthetic extract, sweetness and umami decrease considerably.

2) Alanine serves to produce sweetness, although not a great deal.

3) Glutamic acid contributes greatly to umami. When it is removed, the characteristic taste of crab and the sweet sensation decrease considerably.

4) When arginine is eliminated, the overall taste as well as the crab-like taste becomes weak.

5) Each of GMP and AMP contributes slightly to umami.

6) Omission of CMP produces almost no change in taste. (However, since repeated triangle difference tests consistently showed that a difference between two test solutions at least at the 5% level, is significant, CMP may contribute to the taste of crab, although the panelists could not perceive its presence distinctly.)

7) When sodium ions are omitted, sweetness and umami decrease drastically and the crab-like taste disappears completely.

8) If potassium ions are eliminated, the crab-like taste is retained to some extent, but the taste becomes watery.

9) When chloride ions are removed, the test solution becomes almost tasteless.

10) Removal of phosphate ions causes a slight decrease in sweet and salt sensations as well as umami.

It was also sound in a supplementary test that glycine betaine serves to produce a delicate flavor. Proline, taurine, and TMAO, although their concentrations are remarkably high, contribute little to the taste, as do the other minor components. It was thereby confirmed that a synthetic extract containing the above twelve components could reproduce the crab-like taste, although it is weaker than that of the mixture containing all the constituents listed in Table VII.

From these results, we have depicted a model for the construction of the taste of crab meat. This is shown in Figure 7. The nucleus of the crab taste is produced by a limited number of compounds, such as glycine, glutamic acid, arginine, AMP, GMP, sodium ions, and chloride ions. The characteristic taste of crab meat thus formed is elaborated upon and enhanced by such components as alanine, glycine betaine, potassium ions, and phosphate ions, and possibly by CMP. The other components, though their individual contributions are slight, jointly also serve as taste enhancers.

Conclusion

Among a wide variety of sea foods with a great variety of tastes, the overall taste pictures of only a few have been studied by detailed chemical analysis of their extractive components accompanied by organoleptic tests. Sea foods provide fascinating materials for food chemists who are interested in flavor.

Acknowledgement

The author is indebted to Professor M. Ikawa, Department of Biochemistry, University of New Hampshire, and Drs. K. Yamaguchi and T. Hayashi, Faculty of Agriculture, The University of Tokyo, for their assistance in the preparation of this manuscript.

Literature Cited

1. Oishi, K. New Food Industry, 1963, 10, 1.
2. Hashimoto, Y. in "The Technology of Fish Utilization", (Kreuzer, R., Ed.); Fishing News (Books): London, 1965; p.57.
3. Oishi, K. Bull. Japan. Soc. Sci. Fish., 1969, 35, 761.
4. Konosu, S. Bull. Japan. Soc. Sci. Fish., 1971, 37, 763.
5. Suyama, M. Bull. Japan. Soc. Sci. Fish., 1971, 37, 771.
6. Konosu, S. J. Japan. Soc. Food Sci. Technol., 1973, 20, 432.
7. Suyama, M. Suisangaku Series, 1976, No.13, 68.
8. Konosu, S. J. Fish Sausage, 1976, No.206, 24.
9. Konosu, S.; Watanabe, K.; Shimizu, T. Bull. Japan. Soc. Sci. Fish., 1974, 40, 905.
10. Suyama, M.; Suzuki, H. Bull. Japan. Soc. Sci. Fish., 1975, 41, 787.
11. Konosu, S.; Watanabe, K. Bull. Japan. Soc. Sci. Fish., 1976, 42, 1263.
12. Suyama, M.; Hirano, T.; Okada, N.; Shibuya, T. Bull. Japan. Soc. Sci. Fish., 1977, 43, 535.
13. Konosu, S.; Matsui, T.; Fuke, S.; Kawasaki, I.; Tanaka, H. J. Japan. Soc. Food Nutr., 1978, 31, 597.
14. Jones, N. R. in "Flavor Chemistry Symposium"; Campbell Soup Company: Camden, New Jersey, 1961; p.61.
15. Jones, N. R. in "The Chemistry and Physiology of Flavors", (Schultz, H. W., Ed.); Avi Publ. Co.: Westport, Conn., 1967; Japanese Edition translated by Fujimaki, M. and Ichioka, M.; kenpakusha; Tokyo, 1972; p.253.
16. Spinelli, J. J. Food Sci., 1965, 30, 1063.
17. Simidu, W.; Higashi, T.; Ishikawa, T. Bull. Res. Inst. Food Sci., Kyoto Univ., 1952, No.10, 78.
18. Simidu, W.; Kurokawa, Y.; Ikeda, S. Bull. Res. Inst. Food Sci., Kyoto Univ., 1953, No.12, 40.
19. Endo, K.; Kishimoto, R.; Yamamoto, Y.; Shimizu, Y. Bull. Japan. Soc. Sci. Fish., 1974, 40, 67.
20. Hughes, R. B.: J. Sci. Food Agric., 1964, 15, 293.
21. Konosu, S.; Maeda, Y.; Fujita, T. Bull. Japan. Soc. Sci. Fish., 1960, 26, 45.
22. Kodama, S. J. Tokyo Chem. Soc., 1913, 34, 751.
23. Iimura, K.; Umeda, I. Seasoning Sci., 1959, 7, 17.
24. Aoki, K. J. Agric. Chem. Soc. Japan, 1932, 8, 867.
25. Takagi, I.; Simidu, W. Bull. Japan. Soc. Sci. Fish., 1962, 28, 1192.
26. Konosu, S.; Shibōta, M.; Hashimoto, Y. J. Japan. Soc. Food

Nutr., 1967, 20, 186.
27. Take, T.; Otsuka, H. Mem. Fac. Educ., Niigata Univ., 1966, 8, 75.
28. Konosu, S.; Hashimoto, Y. Unpublished data.
29. Konosu, S.; Maeda, Y. Bull. Japan. Soc. Sci. Fish., 1961, 27, 251.
30. Toi, B.; Ikeda, S.; Matsuno, T. Abstr. Papers, 13th Ann. Meetg. Japan Home Econ. Soc., 1961, p.8.
31. Konosu, S.; Maeda, Y. Bull. Japan. Soc. Sci. Fish., 1961, 27, 251.
32. Endo, K.; Hujita, M.; Simidu, W. Bull. Japan. Soc. Sci. Fish., 1962, 28, 833.
33. Simidu, W.; Hujita, M. Bull. Japan. Soc. Sci. Fish., 1954, 20, 720.
34. Hujita, M. Ph.D. Thesis, Kyoto Univ., 1961.
35. Hujita, M.; Endo, K.; Simidu, W. Mem. Fac. Agric., Kinki Univ., 1972, No.5, 61.
36. Hujita, M.; Endo, K.; Simidu, W. Mem. Fac. Agric., Kinki Univ., 1972, No.5, 70.
37. Take, T.; Honda, R.; Otsuka, H. J. Japan. Soc. Food Nutr., 1964, 17, 268.
38. Take, T.; Yoshimura, Y.; Otsuka, H. J. Home Econ. Japan, 1967, 18, 209.
39. Konosu, S.; Yamaguchi, K.; Hayashi, T. Bull. Japan. Soc. Sci. Fish., 1978, 44, 505.
40. Hayashi, T.; Yamaguchi, K.; Konosu, S. Bull. Japan. Soc. Sci. Fish., 1978, 44, 1362.

RECEIVED April 26, 1979.

[illegible]

27. [illegible] 1967, [illegible]

28. [illegible] Quantum [illegible]

29. [illegible]

30. [illegible]

31. [illegible]

32. [illegible]

33. [illegible]

34. [illegible]

35. [illegible]

36. [illegible]

37. [illegible]

38. [illegible]

39. [illegible]

40. [illegible]

RECEIVED April [illegible]

9

Flavor of Browning Reaction Products

AHMED FAHMY MABROUK

Food Sciences Laboratory, U.S. Army Natick Research and Development Command, Natick, MA 01760

With the exception of fresh vegetables and fruits, most consumed food has to be subjected to processing (boiling, broiling, roasting, canning, baking, concentration, pasteurization,...etc.) to render it edible; to increase its consumer acceptance or to extend its shelf life. Table I lists the consumption of food commodities per capita in the United States (1). About 78% of these comodities are processed, and account for approximately 75% of the American family food budget.

Natural food flavors such as terpenes, hydrocarbons, alcohols, aldehydes, ketones, esters, acids, lactones, amines, sulfur compounds are enzymatically produced in fruits and vegetables. On the contrary, processed food develops its characteristic acceptable flavors from chemical reactions within its components at temperatures far below those at which its major components, i.e., lipids, proteins and carbohydrates pyrolyze. Food flavor precursors responsible for the productivity of volatile flavors are given in Table II.

Aqueous flavor precursors undergo nonenzymic browning during processing, which is the most important flavor producing reaction. Also, at low temperature pyrolysis, i.e., roasting between 100-270°C, these compounds undergo thermal degradation, producing new compounds some of which have desirable organoleptic notes. Some degradation products react further. For example, all purines are decomposed upon heating in acid media at temperatures above 100°C to glycine, formic acid, carbon dioxide and ammonia. Methylated purines give rise to methylamine instead of ammonia.

Proteins and glycoproteins participate in the browning reaction via their free-NH_2 groups. Furthermore, during processing, compounds of these two classes undergo hydrolysis and degradation;thus ammonia, hydrogen sulfide, peptides, amino acids, amines and sugars are produced. A delicious taste peptide which has the following structure: H-Lys-Glc-Asp-Glu-Ser-Leu-Ala-OH was isolated from beef gravy (2). In animal cells, the widely occurring peptides are carnosine, anserine, glutathione, phosphopeptides, lipopeptides and nucleopeptides. The largest class of

Table I

Consumption of Major Food Commodities Per Capita in USA, lb

Commodity	1960	1970	1974	1975	1976	1977
Meats:	134.1	151.4	152,5	145.4	155.3	154.8
Beef	64.3	84.1	86.4	88.9	95.7	93.2
Veal	5.2	2.4	1.9	3.5	3.3	3.2
Lamb and mutton	4.3	2.9	2.0	1.8	1.7	1.5
Pork (excluding lard)	60.3	62.0	62.2	51.2	56.4	56.9
Fish (edibly weight)	10.3	11.8	12.2	12.2	12.9	12.8
Poultry products:						
Eggs	42.4	39.5	36.6	35.4	34.9	34.5
Chicken (ready-to-cook)	27.8	40.5	41.1	40.3	43.3	44.3
Turkey (ready-to-cook)	6.2	8.0	8.9	8.6	9.2	9.2
Dairy products:						
Cheese	8.3	11.5	14.6	14.5	15.9	16.3
Condensed and evaporated milk	13.7	7.1	5.6	5.0	4.7	4.4
Fluid milk and cream (prod.weight)	321.0	296.0	288.0	291.1	292.0	289,4
Ice cream (product weight)	18.3	17.7	17.5	18.7	18.1	17,7
Fats and oils-total, fat content	45.3	53.0	53.2	53.3	56.1	54.4
Butter (actual weight)	7.5	5.3	4.6	4.8	4.4	4.4
Margarine (actual weight)	9.4	11.0	11.3	11.2	12.2	11.6
Lard	7.6	4.7	3.2	4.0	3.6	3.5
Shortening	12.6	17.3	17.0	17.3	18.1	17.5
Other edible fats and oils	11.5	18.2	20.3	20.3	22.0	21.6
Fruits:						
Fresh	90.0	79.1	76.3	81.3	83.6	82.4
Citrus	32.5	27.9	26.8	28.7	28.6	25.2
Noncitrus	57.5	51.2	49.5	52.6	55.0	57.2

Table I (cont'd)

Consumption of Major Food Commodities Per Capita in USA, lb

Commodity	1960	1970	1974	1975	1976	1977
Processed:						
Canned fruit	22.6	23.3	19.6	19.3	19.2	19.5
Canned juice	13.0	14.6	14.7	15.3	15.3	13.7
Frozen (including juices)	9.1	9.8	11.3	12.6	12.2	12.4
Chilled citrus juices	2.1	4.7	5.2	5.7	6.2	5.8
Dried	3.1	2.7	2.5	3.0	2.7	2.6
Vegetables:						
Fresh	96.0	91.0	93.6	93.9	94.5	93.1
Canned, excluding potatoes and sweet potatoes	43.4	51.2	53.3	52.1	52.8	52.8
Frozen, excluding potatoes	7.0	9.6	10.2	9.7	10.2	9.7
Potatoes, (including fresh equivalent of processed)	105.0	115.3	112.3	120.2	114.7	120.7
Sweetpotatoes, (Including fresh equivalent of processed)	6.5	5.2	5.1	5.3	5.3	5.0
Grains:						
Wheat flour	118	110	106	107	111	107
Rice	6.1	6.7	7.6	7.7	7.2	7.8
Other:						
Coffee	11.6	10.5	9.5	9.0	9.4	6.9
Tea	0.6	0.7	0.8	0.8	0.8	.9
Cocoa	2.9	3.1	3.0	2.6	3.0	2.7
Peanuts (shelled)	4.9	5.9	6.4	6.5	6.3	6.5
Dry edible beans	7.3	5.9	6.7	6.5	6.3	6.0
Melons	23.2	21.2	17.2	17.5	18.6	19.0
Sugar (refined)	97.4	101.8	96.6	90.2	94.7	95.7

Reference (1)

Table II

Food Flavor Precursors and Their Flavor Components

Precursors		Possible Flavor Components Produced	
Class	Number	Per Compound	Per Class
a- Aqueous Flavor Precursors			
1- Glycoproteins			
2- Glycopeptides			
3- Proteins			
4- Peptides	60	40	2,400
5- Amino Acids)			
6- Amines)	35	40	1,400
7- Nucleotides)			
8- Nucleotide sugars)	35	40	1,400
9- Nucleotide sugaramine)			
10- Nucleotide acetylsugaramine)			
11- Peptide bound nucleotide	12	40	480
12- Nucleosides	12	40	480
13- Purines and pyrimidines	12	25	300
14- Sugars)			
15- Sugaramine)	50	50	2,500
16- Sugar phosphate)			
17- Organic acids	25		
Subtotal	>241	>275	>8,960
b- Non-Auqeous Flavor Precursors			
1- Neutral lipids	25	50	1,250
2- Polar Lipids	25	50	1,250
3- Isoprenoids	150	5	750
4- Carotenoids	10	25	250
5- Unsaponifiable compounds			
Subtotal	>210		>3,500
22 Grand total	>451		>12,460

peptides in plants consist of γ-glutamyl dipeptides along with some tripeptides, one of which is homoglutathione (γ-glutamylcysteinyl-β-alanine) which replaces glutathione in some enzymic reactions (3). About thirty-five amino acids and amines occur naturally in foods. D-ribose is found in nucleotides, D-ribulose, D-lyxose, D-xylulose, glucose, fructose, sedoheptulose, and glyceraldehyde are present as phosphoric acid esters in the products of carbohydrate metabolism. Fucose is a constitutent of blood group substances which contain galactose and mannose. The presence of carbohydrate alcohols in vegetables, fruits and animals may be accounted for by a reduction of the reducing sugars, or through decarboxylation of the corresponding higher aldonic acids. Sorbose, glucose, sucrose, inositol, mannitol, pentitol and two unidentified sugars were reported in cocoa beans (4).

Sugar acids, mono-, di-, and tri-carboxylic acids, keto-, hydroxy-acids as well as unsaturated organic acids are present in foods. While uronic acids participate in the browning reaction, others might accelerate or retard the reaction.

Non-aqueous flavor precursors contribute directly and indirectly to food flavors. Boar meat lipids contain 5-androst-16-ene-3-one which is responsible for noticeable undesirable notes (5). The characteristic flavors of mutton and goat meat are attributed to the presence of odd-numbered n-fatty acids, and abnormal proportion of branched chain fatty acids (6). During processing of foods, lipids undergo autoxidation, hydrolysis, dehydration, decarboxyltion, and degradation. Thermal degradation of neutral lipids, polar lipids and free fatty acids produce the following classes of compounds (7-13) :-(1)n- and iso-alkanals, (2) alkadianals, (3) oxo-alkanals, (4) alkenals, (5) alkadienals, (6) aromatic aldehydes, (7) methyl ketones, (8) saturated and unsaturated alcohols, (9) γ- and δ- lactones, (10) hydrocarbons, (11) keto- and hydroxy-acids, (12) dicarboxylic acids, and (13) saturated and unsaturated fatty acids, containing less carbon atoms than the parent ones.

Lipids in foods vary from traces as in cereals to 30-50% as in nuts. The physical state and distribution of lipids vary considerably among food items. In each item lipid distributions affect its flavor as it undergoes chemical reactions and act as a flavor components vehicle or partitioning medium. Furthermore, lipids have a pronounced effect upon the structure of food items. Fatty acids of neutral (triglycerides) and polar lipids of beef and pork are tabulated in Table III.

Pork fat contains more unsaturated fatty acids than beef and its linoleic acid content is double that in beef. Heat produced volatiles in red meat fats are listed in Table IV. The presence of pyrazines in the volatiles of beef fat is due to the presence of nitrogenous compounds in the fat amounting to 0.1-0.2%N, (Kjeldhal method). These compounds might be proteins, peptides, amino acids, and amine moities in polar lipids. Such compounds

Table III

Fatty Acids Composition of Beef and Pork Lipids

Fatty Acids	Triglcerides Beef	Triglcerides Pork	Polar Lipids Beef	Polar Lipids Pork
	% of Total Fatty Acid Content			
a.- Saturated				
Capric	0.1	0.1	-	-
Lauric	0.1	0.2	-	-
Myristic	2.2	1.2	2.6	2.0
Palmitic	27.5	23.9	13.2	20.0
Stearic	16.9	11.6	15.6	11.0
Total	46.8	37.0	31.4	33.0
b.- Monounsaturated				
Tetradecenoic	1.0	-	0.9	0.2
Palmitoleic	4.7	7.4	2.2	2.3
Oleic	41.3	45.3	21.2	16.2
Total	47.0	52.6	24.3	18.8
c.- Dienoic				
Tetradecenoic	0.6	-	1.3	0.6
Linoleic	4.4	8.7	20.2	27.9
Docasadienoic	-	-	-	0.9
Total	5.0	8.7	21.5	29.3
d.- Trienoic				
Linolenic	1.1	1.6	1.8	1.0
Eicosatrienoic	-	-	1.9	1.6
Total	1.1	1.6	3.7	2.6
e.- Tetraenoic				
Arachidonic	0.1	0.1	19.1	16.3
Total	0.1	0.1	19.1	16.3

Reference (14)

Table IV

Heat Produced Carbonyls In Red Meat Fats

Volatile Compounds	Pork	Beef	Lamb
ALKANALS			
Ethanal(Acetaldehyde)	10	9a,10	
n-Propanal	10	9a,10	
n-Pentanal	7	8	
n-Hexanal	7,10	8,9,10	10
n-Heptanal	7	8,9	
n-Octanal	7,10	8,9	
n-Nonanal	7,10	8,9	10
n-Decanal	7	8,9	
n-Undecanal	7		
n-Dodecanal	7		
METHYL-ALKANALS			
Methylpropanal		9a	
Methylbutanal	7	8,9a	
ALKANEDIAL			
Ethanedial(Glyoxal)		9a	
OXOALKANAL			
2-Oxopropanal(Pyruvaldehyde)		9a	
2-ALKENAL			
2t-Butenal(Crotonal)		9a	
2t-Heptenal	7	8,9	
2t-Octenal	7,10	8,9	
2t-Nonenal	7,10	8,9	10
2t-Decenal	7,10	8,9	10
2t-Undecenal	10	8	10
2,4-Alkadienal			
2,4-Heptadienal	10		
2,4-Nonadienal	10		
2,4-Decadienal	10	8,10	
2-ALKANONE			
Acetone		10	
2-Butanone		9a	
2-Decanone	7	7	
2-Undecanone		7	
2-Tridecanone		8	
2-Pentadecanone		8	
2-Heptadecanone		8	

Table IV (cont'd)

Heat Produced Carbonyls in Red Meat Fats

Volatile Compounds	Pork	Beef	Lamb
HYDROXY-2-ALKANONE			
3-Hydroxy-2-butanone(acetoin)		9a	
PYRAZINE			
2,5-Dimethyl-		8	
2-Ethyl-		8	
2-Ethyl-3,6-dimethyl-		8	
2-Ethyl-5-methyl-		8	
2,3,5-Trimethyl-		8	

Reference (7,8,9,10)
"a" designates heating in nitrogen atmosphere, otherwise in air.

degrade during heating and produce ammmonia and basic compounds. When ammonia, amino acids and degraded nitrogenous compounds react with α-dicarbonyls produced from lipid oxidation, pyrazines are formed.

Cooked chicken flavor concentrate (13) contained the following aldehydes; 3c-nonenal; 4c-decenal; 2t,4c-decatrienal; 2t,5c-undecadienal; 2t-dodecenal; 2t,4c-docadienal; 2t,6c- and 2t,6t-dodecadienal 2t-tridecenal; 2t,4c-tridecandienal; 2t,4c,7c-tridecatrienal; and 2t,4c-tetradecadienal. Three of these aldehydes: 4c-decenal; 2t,6c-dodecadienal; and 2t,4c,7c-tridecatrienal are typical breakdown of arachidonic acid, and to a major extent also 2t,5c-undecadienal. These aldehydes play an important role in cooked chicken flavor via the browning reaction. 2,4-Decadienal which is considered to be a key compound of chicken aroma, was found in the volatiles of cooked chicken (15,16). 2,4-Heptadienal; 2,4-nonadienal and 2,4-decadienal were identified in heated pork fat (10) and heated beef fat (8,10). Dicarbonyls, dienals, trienals undergo amino-carbonyl reactions, resulting probably in the formation of meaty flavor notes in specific proportions characteristic of each species. Although their yield is small in comparison with monocarbonyls, their high reactivities are responsible for thier important role in flavor production.

The complexity of food flavor precursors is manifested by the number of compounds estimated in each class of compounds and by the possible total number in a single raw food. The possible number of flavor components produced per one compound of flavor precursor is estimated in the range 10-150, but is reduced to 5-40 to eliminate compounds that might be produced by more than one precursor and to account for variations in processing methods. One hundred compounds were identified in thermal degradation products of glucose (17,18,19): aldehydes ketones, aromatics, furans, oxygenated furans, non-volatiles. Sugars react with ammonia at temperatures ranging from 20 to 260°C., to produce heterocyclic compounds : substituted imidazoles, substituted pyrazines, substituted piperazines, pyridine and substituted pyridines(20). At 0-120°C., α-dicarbonyls (2,3-butadione, ethanedial "glyoxal", 2-oxopropanal "pyruvaldehyde", and α-hydroxycarbonyl "glycolaldehyde" react with ammonia in presence of formaldehyde to give substituted imidazoles. Dipeptides react with sugars either as an entity or as amino acids produced by hydrolysis during processing. In the first case, one compound is formed, pyrazinone (21), in the second case numerous flavor compounds are produced from the reaction of the two amino acids with sugar. The reactivity of dipeptides towards reaction with carbonyl compounds is much higher than that of amino acids. The taste of glycyl-L-leucine is very bitter, the product of its reaction with glyoxal has an astringent, a little sour and later a mild taste. Some products from the browning reaction and pyrolysis of flavor precursors react with ammonia and hydrogen sulfide, thus increasing the number of flavor components produced from food precursors. The molecular structure of the amino acid influences the flavor notes in the food. For example, thiophenes are reported in the volatiles produced from the pyrolysis of cysteine or from its reaction with glucose or pyruvaldehyde (22,24) but were absent in the case of cystine (23,24).

Hundreds of compounds have been identified in the volatile flavor components of processed foods. Hydrocarbons, alcohols, ethers, aldehydes, ketones, acids, acid anhydrides, esters, aromatic, lactones, pyrones, furans, pyridines, pyrroles, n-alkylpyrrole-2-aldehydes, pyrazines, sulfides, disulfides, thiols, thiophenes, thiazoles, trithiolanes, thialdine ...etc. Each compound has a characteristic note and a specific threshold which makes its contribution to food flavor unique. But none of the individual compounds were reported to completely produce the characteristic flavor of the processed food. This indicates that the food flavors are of mixed flavor notes of numerous compounds, while others are necessary for its synergistic effect, others exert a background of modifying effect.

Browning reaction has become a major topic of food research because of its relevance to the desirable and undesirable changes (flavor, color, texture, and nutritive value) occurring in foods during processing and storage. During the Second World War and its post era, research on food deterioration due to nonenzymic browning received considerable attention due to the problems resulting from the necessity to prepare dehydrated foods and food concentrates which had to be stored under tropical conditions without serious deterioration.

The purpose of this paper is to review the numerous papers published on flavors, tastes and odors resulting from the browning reaction. Investigations of model systems which have been observed under laboratory conditions are considered and their possible significance in basic and industrial processes will be discussed. Speculation on the possible correlation between model system results and specific processed food items will be presented. Results of recent work in our laboratory on flavor notes developed upon heating ribose with various amino acids will be discussed.

The numerous purely chemical papers are considered to be outside the scope of this paper. The reader is referred to several reviews on the subject (25-30). Comprehensive reviews on nonenzymic browning in relation to specific food problems had been published (31,32,33).

As flavor production in natural food is governed by too complicated reactions due to its complex components (Table II), chemists concentrated their research efforts on simpler systems to understand the reactions involved and their products.

Amino Acids-Sugars Model Systems

Model system studies of the reaction between single amino acids and naturally occurring substances capable of reacting with them has furnished valuable information leading toward an understanding of food flavors. The well known Strecker degradation of α-amino acids, Schonberg and Moubacher (25), has been used as the central reaction around which other amino compound degradation systems may be oriented. In this reaction α-amino acids are deaminated and decarboxylated by specific carbonyl compounds or others to yield aldehydes and ketones containing one carbon atom less. β-amino acids also undergo oxidative deamination and decarboxylation to a ketone with one less carbon atom; e.g., β-amino-n-butyric acid produces 2-propanone, which also results from the Strecker degradation of α-amino-isobutyric acid. Organic di- and tricarbonyls are not the only oxidants effective in Strecker degradations. Hydrogen peroxide in presence of ferrous sulfate degrade α-amino acids at room temperature ; for example glycine yields formaldehyde. The co-oxidation of sulfur-containing amino acids in an auto-oxidizing lipid system indicates that lipid peroxides act as oxidants (34). The resulting carbonyl compounds from amino acids through Strecker degradation participate in the formation of desirable characteristic flavors during processing or undersirable ones responsible for flavor deterioration during storage.

Flavors and odors given by amino acids and sugars in dilute aqueous solutions at different temperatures has been the subject of intensive studies by various researchers. Tables V and VI summarize the descriptive aroma evolved from reacting carbonyl compounds with amino acids at 100, and 120/180°C., respectively

Table V

Aromas Developed by the Reactions of Carbonyl Compounds and Various Amino Compounds at 100°C

Amino Compounds	Dihydroxyacetone	Glucose	Fructose	Maltose	Sucrose
Glycine	baked potato(35)	caramelized sugar (36,39), faint beer(40).	unpleasant carmel smell(36).	weak(36)	objectionable weak NH_3(36)
α-Alanine	weak caramel(35)	beer aroma(40)	--	--	--
Valine	strong, yeasty protein hydrolyzate(35)	rye bread(37) fruity, aromatic (39)	--	--	--
α-Aminobutyric	--	maple(38)	--	--	--
Leucine	strong, cheesy,	sweet chocolate(37) toast(39), bread(40) rye bread(38)			
Isoleucine	moderate crust (35)	musty(37), fruity, aromatic(39)	--	--	---
Serine	vaguely bread-like(35)	maple syrup(38)	--	--	--
Threonine	very weak(35)	chocolate(37) maple(38)	--	--	--
Phenylglycine	--	bitter almond(38)	--	--	--
Methionine	baked potato(35)	overcooked sweet potato(36) potato(37)	objectionable chopped cabbage(36)	overcooked cabbage(36)	unpleasant burned wood(36)

Table V (cont'd)

Aromas Developed by the Reactions of Carbonyl Compounds and Various Amino Compounds at 100°C

Amino Compounds	Dihydroxyacetone	Glucose	Fructose	Maltose	Sucrose
Cysteine	mercaptan , H_2S (35)	meat(39) sulfide(37)	--	--	--
Cystine	--	meat, burnt turkey skin(39)	--	--	--
Proline	very strong cracker,crust, toast(35)	corn-like(39) burnt protein (37)	--	--	--
Hydroxyproline	weak, vaguely like proline(35)	potato(39)	--	--	--
Arginine	very weak(35)	popcorn(37) butterscotch(39)	--	--	--
Histidine	very weak(35)	buttery note(39), none(37)	--	--	--
Glutamine	--	chocolate(37)	--	--	--
Methylamine	--	empyreumatic taste(40)	--	--	--
Ammonia	--	tarry odor, bitter taste(40)	--	--	--
Phenylalanine	very strong, hyacinth(35)	rancid caramel, unpleasant(36) violets(37) rose perfume(39)	stinging smell, very objectionable(36)	pleasant sweet caramel(36)	unpleasant sweet caramel(36)

Table V (cont'd)

Aromas Developed by the Reactions of Carbonyl Compounds ans Various Amino Compounds at 100°C

Amino Compounds	Dihydroxyacetone	Glucose	Fructose	Maltose	Sucrose
Tyrosine	--	caramel(37)	--	--	--
Aspartic	very weak(35)	rock candy(37)	--	--	--
Glutamic	chicken broth(35)	oldwood pleasant(36)	too weak(36)	too weak(36)	pleasant caramel(36)
Arginine	--	buttery note(39)	--	-	--
Lysine	strong dark corn syrup(35)	baked sweet potato(36)	objectionable fried butter (36)	unpleasant wet wood (36)	rotten wet potato(36
α-Methylamino-butyric	--	maple(38)	--	--	--
α-Amino-isobutyric	--	maple(38)	--	--	--

References (35,40)

Table VI

Aromas Developed by Heating Glucose with Various Amino Acids at 120°C and 180°C

Amino Acid	Aroma Description 120°C	Aroma Description 180°C	Amino Acid	Aroma Description 120°C	Aroma Description 180°C
Valine	moderate breadcrust(41)	penetrating chocolate(37)	Lysine	weak(41)	bread-like(37)
Leucine	moderate breadcrust(41)	burnt cheese(37)	Methionine	strong baked potato(41)	potato(37)
Isoleucine	moderate breadcrust(41)	objectionable burnt cheese(37)	Proline	strong bread-crust cracker (41)	pleasant bakery aroma(37)
Threonine	weak(41)	burnt(37)	Histidine	no significant aroma(41)	cornbread(37) buttery note(38)
Phenylalanine	strong flower (41)	violets,lilac (41)	Glutamine		butterscotch(37)
Aspartic	strong bread-crust(41)	caramel(37)	Tryptophan	strong(41)	
Glutamic	moderate(41)	brunt sugar(37)			
Arginine	no significant aroma(41)	brunt sugar(37)			

References (37,38,41)

(35-41). The aroma produced by the interaction of sugars and amino acids ranged from pleasant to objectionable. The molecular structure of sugar influences the resulting aroma of the products of its reaction with the amino acid. While a pleasant caramel sweet aroma developed upon reacting phenylalanine with maltose, an unpleasant caramel aroma developed with fructose, and a hyacinth aroma in case of dihydroxyacetone. While the aroma of methionine-sugar browning product was reminiscent of a pleasant baked potato in presence of dihydroxyacetone, aroma notes of overcooked potatoes developed with glucose. In presence of a reducing disaccharide (maltose), overcooked cabbage aroma was noticed and an unpleasant burnt wood aroma developed in presence of non-reducing disaccharide (sucrose). Proline, valine and isoleucine in presence of glucose gave a pleasant bakery aroma (41). Keeney and Day (42) studied the character odor of the products of Strecker Degradation of amino acids using isatin as the oxidant (Table VII).

Table VII

Aroma of Strecker Degradation Products of α-Amino Acids with Isatin (Diketodihydroindole)

Amino Acids	Aldehyde	Sensory Description Keeney & Day(42)	Sensory Description Others(43,44)
α-Alanine	acetaldehyde	malty	reminiscent of wine, coffee(43)
Valine	2-methyl-propanal	apple	fruity,banana-like(43) green,pungent sweet(44)
Leucine	3-methyl-butanal	malty	fruity,peach-like(43) burnt,green,sickly(44)
Isoleucine	2-methyl-butanal	malty-apple	burnt,sickly(44)
Norleucine	pentanal	flower	slightly fruity,herb-aceous,nut-like(43) burnt-green(44)
Phenyl-alanine	phenylacet-aldehyde	violets	hyacinth odor,resemble benzaldehyde in taste(43)
Glutamic		bacterial agar	
Cystine	methional	cheesy-brothy	

References (42,43,44)

There is some agreement and disagreement between the sensory notes stated in Tables V and VII, which may be attributed to variations in the concentration of the reactants and lack of agreement on description of sensory notes. The discrepancies can be resolved, by a careful consideration of the specific conditions of the reaction. It is possible that both the nature and extent of the reaction vary widely with the principal variables (concentration, temperature, heating period, specific reactants). Unfortunately, many of the papers insufficiently define the experimental conditions for an adequate evaluation of the results. The description of odor and taste notes of a product or a compound varies according to its concentration, the media used for evaluation (water, paraffin oil, skim milk...etc), and the sensory evaluator.

The data in Table VI indicate that the aromas developed when sugars and amino acids were heating at 120 and 180°C, were quite different from those produced at 100°C. Noticeable differences exist between aroma notes developed at 120°C and 180°C, this may be attributed to sugar degradation notes which become noticeable in the system heated at 135°C and above.

All aliphatic compounds containing primary or secondary amine group react in browning reaction but at different rates depending upon the molecular structure. The reactivity of the reactants (amino acids, carbonyl compounds) in the browning reaction had been reported in the literature as:

a. intensity of color developed,

b. amount of carbonyl compounds determined by GC or as dinitrophenylhydrazones.

c. percent loss of reactants (one reactant or both).

It is very hard to compare results reported in the literature, as there are various variables influencing the data. Concentration of reactants, ratio of amine compound to sugar or carbonyl, reaction in buffer or water, molarity and type of buffer used, pH of buffer, duration of reaction. Rate of carbonyl compounds formation will be the criterion used for reactivity except in few occasions where the percent loss of one of the reactants or both will be referred to in the manuscript.

Rooney et al (45) reported that the rate of carbonyl formation varied with the molecular structure of sugar. Xylose was most reactive as it produced the greatest quantity of carbonyls, followed by glucose, then maltose. In the presence of these sugars isoleucine was more reactive than phenylalanine. In a study on the Strecker degradation of valine-carbonyl, diacetyl showed the greatest reactivity followed by sorbose>arabinose>xylose>fructose>glucose>sucrose>rhamnose, Self(46). The number of volatile carbonyls produced by the reaction of glucose and glutamic at pH 5.0,6.5 and 8.0 at 100°C was almost equal. When these systems were heated at 180°C, the number of volatiles was greater and increased more drastically as pH increased to alkalinity (36). The amount of aldehydes produced from mixtures of 0.01M amino acid and 0.1M glucose in water was far below those produced in 0.01M phosphate buffer at pH6.5, which in turn was far below the amount obtained at pH 7.5. The addition of phosphate buffer increased the amount of aldehydes produced signigicantly. The amount of aldehyde produced is also influenced by the chain length of the acid. In general, straingt chain amino acids produced more aldehydes than branched chain amino acids of the same number of carbon atoms (46).

Alanine, valine, serine, glutamic, glutamine, methionine, taurine, histidine, creatine, citrulline, carnosine, cystine, systeine, aspartic, asparagine, leucine, isoleucine, tyrosine, phenylalinine, tryptophan, α-aminobutyric, proline, ornithine, 2-pyrrolidone-5-carboxylic, threonine, glycine, lysine, arginine, cysteine were identified in aqueous red meat flavor precursors (47). Ribose was also reported as the major sugar. Flavor evaluation of the browning reaction products of these amino acids and ribose were undertaken. 0.1M solution of each amino acid was prepared by dissolving the amino acid in 0.05M Sørensen phosphate buffer and readjusting the pH by the addition of monopotassium or dipotassium phosphate. 0.1M ribose solution was prepared in the same buffer. Ten ml of amino acid and 10ml of ribose solutions were pipetted in 50ml glass ampoule and 50ml round bottom flask with 24/40 glass joint. 1.0 mMoles of insoluble amino acids were weighed directly in reaction vessels, and 10ml buffer added. To study the effect of heating on flavor of amino acids in absence of ribose, 10ml buffer were pipetted in glass ampoules and flasks along with the amino acid. The ampoules were sealed after evacuation, and heated for one hour at 180°C. The flasks were heated in glycerol bath at 100°C, after fitting with condensors. After the heating period, the flasks and the ampoules were removed, cooled in wet ice, and the contents transferred to vials and kept at -10°C. All experiments were run in triplicates. In absence of ribose, while sulfur containing amino acids solutions produced sulfury notes upon heating, none of the others developed any flavor notes. Table VIII lists the striking flavor notes produced by heating various amino acids-ribose solutions at 100° and 180°C.

Table VIII

Flavor Notes Produced by Heating Amino Acid-Ribose AT 100°C and 180°C

Amino Acid	Flavor Description 100°C	180°C
Alanine	very mild caramel	sweet burnt caramelized sugar
Valine	sickly sweet	penetrating burnt chocolate
Serine	sweet bouillion	burnt sugar
Glutamic	brothy, slightly sweet, lingering in mouth	roasted meat
Methionine	sulfury, savory	crust of roast meat
Cysteine	sulfury, rotten egg	sulfury, spicy meat
Taurine	pleasant toffee	sickly, sweet, burnt caramel

Table VIII (cont'd)

Flavor Notes Produced by Heating Amino Acid-Ribose At 100°C and 180°C

Amino Acid	Flavor Description 100°C	180°C
Tryptophan	oily aromatic, sugar sweet	oily aromatic, naphththalene
Phenylalanine	sharp flower	flowery with aromatic and caramel notes, undersirable
Histidine	salty, slightly bitter caramelized toffee	pleasant slightly burnt caramel
Creatine	slightly salty	slightly sweet caramel
Tyrosine	slight caramel	custard slightly burnt sugar
Leucine	bitter almond	toasted bread
Aspartic	bread crumb	caramelized bread crust
Cystine	hard boiled egg yolk	meaty with H_2S note
Glutamine	caramel with burnt sugar note	butterscotch
Asparagine	desirable burnt sugar	creamy butterscotch sugar
α-Aminobutyric	undersirable burnt sugar	maple
γ-Aminobuturic	burnt sugar	maple
β-Aminobutyric	custard	maple
Arginine	burnt sugar	buttery, burnt sugar
Ornithine	bread crumb	bread-like
Proline	bread crumb	cracker, toast
Cysteic	undersirable, sulfury	meaty, sulfury
2-Pyrrolidone -5-carboxylic	brothy	meaty, pleasant
Isoleucine	aromatic, undesirable	burnt cheese
Glycine	caramel, faint	burnt sugar
Lysine	custard	bread
Homocystine	canned milk	scorched boiled milk
Threonine	custard weak	burnt custard
Citrulline	toffee-like	meaty
Carnosine	buttery-toffee	meaty

The previous discussion covered Strecker degradation of amino acids sugar aqueous solutions. A limited number of investigations were carried out on the nature of the browning reaction at high temperature, i.e. low temperature pyrolysis and at low moisture content. Rohan and coworkers had provided an extensive research on model systems simulating cocoa nib roasting in absence of lipid. Amino acids and sugars were responsible for chocolate flavor development (48,49,50,51). While the degradation of amino acids in cocoa beans during roasting was incomplete, the degrading agent, reducing sugar was completely destroyed.

In a study on model systems prepared from single amino acid and glucose in molar ratio 2:1, approximating the composition of shell-free cocoa beans, Rohan and Stewart (52) reported that amino acid destruction from heating was temperature dependent and practically ceased after one hour. Sugar destruction contintued at a rate dependent on reaction temperature until the end of the experiment (Table IX).

Table IX

Destruction of Amino Acids and Peducing Sugars on Heating Chocolate Aroma Precursor Extract

Heating		Destruction Percent			
°C	Min.	Amino Acids	Red. Sugars	Total Sugars	Aroma
50	30	6.2			
50	60	8.9	27.0		
50	120	8.9	41.1		
100	30	18.8	27.5	35.8	Cocoa
100	60	29.6	56.6	79.6	Cocoa
100	120	30.9	81.4	22.8	Cocoa
120	30	43.0	84.1	38.9	Cocoa
120	60	54.4	100	79.2	Cocoa
120	120	54.4	93.5	91.2	Cocoa
150	10	58.8			Cocoa
150	30	81.7			Cocoa
150	60	85.4			
150	120	85.1			

Reference (52) Journal of Food Science

The failure of amino acid degradation to go to completion and the relation between maximal destruction and temperature might be attributed to the fact that individual amino acids reacted at different rates at different temperatures. For example certain amino acids, leucine, arginine, methionine, and lysine showed little or no detectable reaction at 100°C after one hour, while others were reactive at this temperature(Table X). At higher temperatures, rate of amino acid degradation was measurable and proportional to the temperature of the reaction.

The effect of increasing temperature on the amount of amino acid destroyed in one hour was noticeably greater at temperatures above 140°C than below 120°C. Between 120° and 140°C., increase in reaction temperature had less marked effect on the reaction due to changes in the physical condition of the reaction mixture. At about 150°C. glucose melts and , as it might be expected, its mixture with amino acids would begin to melt at lower temperature.

Table X

Destruction of Amino Acids After Heating With Glucose at Different Temperatures for One Hour

Temp. °C	Al	PhAl	Threo	Glu	Meth	Leu	Val	Arg	Lys
	Destruction Percent								
80					0.9		7.0		
90		13.9	2.0	11.1	0.9		14.7	1.5	
100		21.0	22.9	23.1			23.1	1.0	4.5
110	11.3	22.2	41.2	28.8	28.5	10.5	26.2	14.5	19.5
120	24.3	23.2	42.2	26.2	40.5	19.5	28.2	29.0	28.0
130	33.4	32.0	59.4	39.0	44.3	27.2	30.2	33.8	30.2
135			90.0			39.9	42.5	40.0	44.5
140	44.8	59.8		50.9	51.9				
150	76.9			62.1			51.5		

Reference (<u>52</u>)

Journal of Food Science

Table XI indicates that the degree of degradation was influenced by the molecular structure of the amino acid. At 120°C the rate of destruction of amino acids by class was in the following order: sulfur-containing, hydroxy>acidic>neutral, basic, aromatic.

Table XI

Degradation of Amino Compounds After Heating for One Hour in Presence of Glucose

Class of Amino Compounds	Percent Degradation at °C 120	135	150
Neutral			
a- n-Alkyl	20-28	40	45
b- iso-alkyl	-----	42.5	--
Hydroxy	42	90	--
Aromatic	23	--	60
Acidic	36	--	51
Sulfur-containing	41	--	52
Basic	20-29	40-45	--

Reference (<u>52</u>)

Moisture content of the reaction mixture influenced the degradation of amino acids. Rohan and Stewart (52) reported that chocolate precursor aroma extracts when heated in the dry state for one hour at 100°C., lost 30% of the amino acids by degradation. When very lightly moistened the amino acids degradation dropped to 9% under the same reaction conditions. When the reaction mixture was wetted with an equal weight of water, no amino acid degradation was noticed. Cocoa beans moisture content averaged 5 to 6 % which might be sufficient to inhibit browning reaction at lower temperatures, thus in the early stages of roasting the removal of moisture is important.

Roasting cocoa beans results in the production of volatile and non-volatile compounds which contribute to the total flavor complex. 5-Methyl-2-phenyl-2-hexenal, which exhibited a deep bitter persistant cocoa note, was reported in the volatile fraction (53). It was postulated to be the result of aldol condensation of phenylacetaldehyde and isovaleraldehyde with the subsequent loss of water. The two aldehydes were the principal products of Strecker degradation products of phenylalanine and leucine, respectively. Non-volatiles contained diketopiperazines (dipeptide anhydride) which interact with theobromine and develop the typical bitterness of cocoa (54). Theobromine has a relatively stable metallic bitterness, but cocoa bitterness is rapidly noticed and disappears quickly. While cocoa bitterness is felt in the whole mouth, theobromine is recognized by the hind part of the tongue. Furthermore, cocoa bitterness is more intense than that of concentrated aqueous solution of theobromine. The following diketopiperazines had been identified as the components responsible for cocoa bitterness:- cyclo(-Pro-Leu-), cyclo(-Val-Phe-), cyclo(-Pro-Phe-), cyclo(-Pro-Gly-), cyclo(-Ala-Val-), cyclo(-Ala-Gly-), cyclo-(Ala-Phe-), cyclo(-Phe-Gly-), cyclo(-Pro-Asn-), and cyclo-(-Asn-Phe-). Those containing phenylalanine exhibited bitterness resembling that of theobromine. Diketopiperazines have stronger bitterness than the corresponding dipeptides or its two individual amino acids. They develop upon heating proteins to temperature above 100°C. Kato et al (55) reported that roasting serine at 280°C , for 30 min under slow nitrogen stream, resulted in the production of 2,5-diketo-3,6-dimethylpiperazine, which upon alkaline hydrolysis produced the dipeptide alanylalanine. A bitter peptide, cyclo(-L-Leu-L-Trp-), was isolated from casein enzymic digest (56). The formation of cyclo(-Pro-Leu-) in aged sake is responsible for its bitterness (57). Recently, five bitter L-proline-containing diketopiperazines were reported in roasted malt (210°C.) used in dark beer brewing. These compounds and their approximate "bitter threshold % values" in aqueous solutions were: cyclo(-L-Phe-L-Pro-),0.1; cyclo((-L-Leu-L-Pro), 0.05; cyclo(-L-Pro-L-Pro-),0.1; cyclo(-L-Val-L-Pro-),0.1; and cyclo(-L-Ile-L-Pro-),0.05. The authors concluded that these diketopiperazines did not contribute directly to beer bitterness and questioned their role in influencing the taste of the product (58).

More than 300 compounds had been identified in cocoa volatiles, 10% of which were carbonyl compounds (59,60). Acetaldehyde, 2-methylpropanal, 3-methylbutanal, 2-methylbutanal, phenylacetaldhyde and propanal were products of Strecker degradation of alanine, valine, leucine, isoleucine, phenyl-acetaldehyde, and α-aminobutyric acid, respectively. Eckey (61) reported that raw cocoa beans contain about 50-55% fats, which consisted of palmitic (26.2%), stearic (34.4%), oleic (37.3%), and linoleic (2.1%) acids. During roasting cocoa beans these acids were oxidized and the following carbonyl compounds might be produced:- oleic : 2-propenal, butanal, valeraldehyde, hexanal, heptanal, octanal, nonanal, decanal, and 2-alkenals of C_8 to C_{11}. Linoleic : ethanal, propanal, pentanal, hexanal, 2-alkenals of C_3 to C_{10}, 2,4-alkadienals of C_9 to C_{11}, methyl ethyl ketone and hexen-1,6-dial. Carbonyl compounds play a major role in the formation of cocoa flavor components.

Another example of food in which the browning reaction occurs at near anhydrous condition is coffee. Coffee beans are usually roasted at temperatures ranging from 180° to 260°C. for specific period to obtain the desired degree of roasting (light, medium, and dark). Protein, sucrose, and chlorogenic acid were the compounds drastically destroyed in coffee beans upon roasting, Table XII "the data in this table were not corrected for dry weight loss which varied from 2 to 5%"(62).

Table XII Composition of Green and Roasted Coffee

Constituent	Percent, Dry Basis	
	Green	Roasted
Hemicelluloses	23.0	24.0
Cellulose	12.7	13.2
Lignin	5.6	5.8
Fat	11.4	11.9
Caffeine	1.2	1.3
Sucrose	7.3	0.3
Chlorogenic acid	7.6	3.5
Protein(Based on nonalkaloid nitrogen)	11.6	3.1
Trigonelline	1.1	0.7
Reducing sugars	0.7	0.5
Unknown	14.0	31.7
Total	100.0	100.00

Reference (62)

Journal of Agricultural and Food Chemistry

The degree of roasting influenced the magnitude of degradation of sucrose, the major sugar present in coffee, Table XIII.

Table XIII
Effect of Roasting On Sucrose Content In Coffee
(Percent Dry Basis)

	Colombian		Santos	
	Sucrose Content	% Loss	Sucrose Content	% Loss
Green	4.59		5.47	
Light Roast	0.45	90.20	0.68	87.57
Medium Roast	0.17	96.30	0.27	95.06
Dark Roast	0.06	98.69	0.10	98.17

Reference (62)

Feldman et al (62) found no noticeable changes in the contents of the following amino acids in roasted coffee beans proteins: alanine, glutamic, glycine, isoleucine, leucine, phenylalanine, proline, tyrosine, and valine. These findings are in agreement with those reported by Underwood and Deatherage (63). Table XIV showed the degree of destruction of amino acids in coffee bean after roasting. It is quite obvious that the magnitude of nonenzymic browning of amino acids was influenced by its molecular structure and degree of roasting.

Coffee oil contains about 47% linoleic, 8% oleic, 1% hexadecenoic, 32% palmitic, 8% stearic, and 5% behenic and longer chain fatty acids (64). As linoleic acid is the major unsaturated fatty acid in coffee oil, its major oxidation products 2,4-alkadienals and hexen-1,6-dial would play a major role in volatile production. Green coffee beans contain 50 to 60% carbohydrates: 18% nonhydrolyzable cellulose, 13% hydrolyzable cellulose, 13% starches and pectins easily solublized, and 9-12% soluble carbohydrates of which sucrose is the major component. Raffinose and stachyose are the tri- and tetrasaccharides reported in robusta coffee beans. Arabinogalactan and galactomannon are the water-soluble polysaccharides reported in coffee beans (62). The above mentioned carbohydrates, amino acids, lipids along with other flavor precursors produce several hundreds of volatile and nonvolatile compounds through different reactions during roasting. Molecular structures, quantities, and ratios of these compounds influence coffee flavor. The variety of coffee as well as the degree of roasting exert characteristic flavors. The molecular structure of carbonyls influence the type of pryazine formed. While refluxing rhamnose (100g) and ammonia (28%; 40ml) in water (100g) produced methyl- and ethyl-substituted pyrazines, glucose and ammonia reaction resulted in formation of methyl-substituted

Table XIV
Composition of Amino Acids in Green and Roasted Coffee
(After Acid Hydrolysis),%

Amino Acid	Haita			Columbia			Angola Robusta		
	Green	Roast I	Roast II	Green	Roast I	Roast II	Green	Roast I	Roast II
Arginine	4.72	0.00	0.00	3.61	0.00	0.00	2.28	0.00	0.00
Asparagine	10.50	9.07	9.02	10.61	9.53	7.13	9.44	8.94	8.19
Cysteine	3.44	0.38	0.34	2.89	0.76	0.69	3.87	0.14	0.14
Histidine	2.85	1.99	2.17	2.79	2.27	1.61	1.79	2.23	0.85
Lysine	6.19	2.54	2.74	6.81	3.46	2.76	5.36	2.23	2.56
Methionine	2.06	2.32	1.48	1.44	1.08	1.26	1.29	1.68	1.71
Serine	5.60	1.77	1.26	5.88	2.60	0.80	4.97	0.14	0.00
Threonine	3.73	2.43	1.83	3.82	2.71	1.38	3.48	2.37	1.08
Total	39.09	20.50	18.84	37.85	22.41	15.63	32.48	17.73	14.53
Loss %		47.56	51.80		40.79	60.02		45.41	55.26

Reference (62)

Journal of Agricultural and Food Chemistry

pyrazines only (53). The flavors of the products of cysteine-glucose and cysteine-pyruvaldehyde in anhydrous condition at different temperatures 80-190°C. are listed in Table XV (24). Five-mM cysteine+5-mM glucose or pyruvaldehyde reacted at different temperatures for 5 minutes and then evaluated by flavor panel.

Table XV

Flavors Produced by Cysteine-glucose And Cysteine-pyruvaldehyde at Different Temperatures

Reactants	Flavor Description, at °C.				
	80°	100°	130°	160°	190°
Cysteine & Pyruvaldehyde	Japanese rice cracker weak.	Sesame, weak	Sesame.	Japanese rice cracker with sesame-like.	Sesame-burnt.
Cysteine & Glucose	no odor	no odor	Japanese rice cracker with seasame-like, sweet.	Japanese rice cracker with sesame-like.	Sweet sesame, burnt.

Reference (24)

Agricultural and Biological Chemistry

Pyruvaldehyde is a liquid at room temperature and boils at 72°C, thus when cysteine-pyruvaldehyde mixture was heated at 80°C, the components are in solution and flavor notes reminiscent of Japanese rice cracker developed. As reaction temperatures increased gradually other flavor notes developed. In the case of cysteine-glucose system, no reaction took place until the reaction temperature reached 130°C. The flavor of cysteine-glucose was comparable to that of cysteine-pyruvaldehde at 160°C, with one exception, the glucose system had a sweet note. As temperature increased the flavor impression of both systems increased in similarity. The volatile compounds produced at 160°C in the presence of pyruvaldehyde were different from those in presence of glucose. While thiazole and thazolines were absent in the volatiles of cysteine-glucose, cysteine-pyruvaldehyde volatiles were devoid of pyridines, picolines and furans (24).

Food flavors consist of numerous compounds, none of which alone is characteristic of specific food. Classes of compounds which emcompass food flavors are:- hydrocarbons (aliphatic, alicyclic, aromatic); carbonyls (aldehydes, ketones); carboxylic acids, esters, imides, anhydrides; alcohols, phenols, ethers; alkylamines, alkylimines; aliphatic sulfur compounds (thiols, mono-, di- and tri-sulfides); nitrogen heterocyclics (pyrroles, pyrazines, pyridines); sulfur heterocylics (thiophenes, thiazoles, trithiolane, thialidine); and oxygen-heterocyclics (lactone, pyrone, furan). Discussion will be limited to striking developments in heterocyclics.

Heterocyclic Compounds

During browning reaction in foods, heterocyclic compounds of known five- and six-membered ring systems containing one or more atoms than carbon as ring members, are produced. These compounds encompass several classes of compounds that exhibit desirable characteristic organoleptic notes, i.e. toasted, bread-like, roasted, brothy, mushroom-like, nutty, ... etc., or undesirable notes, e.g. peppery ammoniacal, obnoxious, ... etc. The nomenclatures and the molecular formulas of heterocyclic compounds that most frequently encountered in food volatiles are given in Figures 1 and 2.

Nitrogen-heterocyclic

Pyrrole and pyrrole derivatives. The chemical and biological value of pyrrole and its derivatives cannot be overemphasized, natural pigments, heme, chlorophyll, bile pigments and enzymes like cytochromes, contain pyrrole nucleus. Also, many alkaloids, proline and hydroxyproline contain the reduced pyrrole ring (pyrrolidine). Pyrrole and its derivatives are found among the products of the browning reaction products in processed foods. Alkylpyrroles have intense petroleum-like odor, but they give sweet, slightly burnt aroma on extreme dilution. On the other hand acylpyrroles have characteristic sweet smoky, and a little medicine-like odor (65). Although alkyl- and acyl- pyrroles do not exhibit favorable aroma like the desirable roasty aroma of pyrazines, they may play an important role in the characteristic roasty flavor of processed foods. In anhydrous condition, N-acetonylpyrrole, which had a bread-like aroma was isolated from the roasting products of proline-glucose, hydroxyproline-glucose, and pyrrolidine-pyruvaldehyde (66). 1-Pyrroline which exhibited an amine like or corn-like odor was the product of Strecker degradation of proline-, and ornithione-sugar or- polycarbonyl reagents in aqueous solutions (67). 2-Pyrrolealdehyde, 2-acetylpyrrole, 2-propionylpyrrole, N-methyl-2-pyrrolealdehyde, N-methyl-2-acetyl pyrrole, 5-methyl-2-pyrrolealdehyde, N-methyl-5-methyl-2-pyrrole-

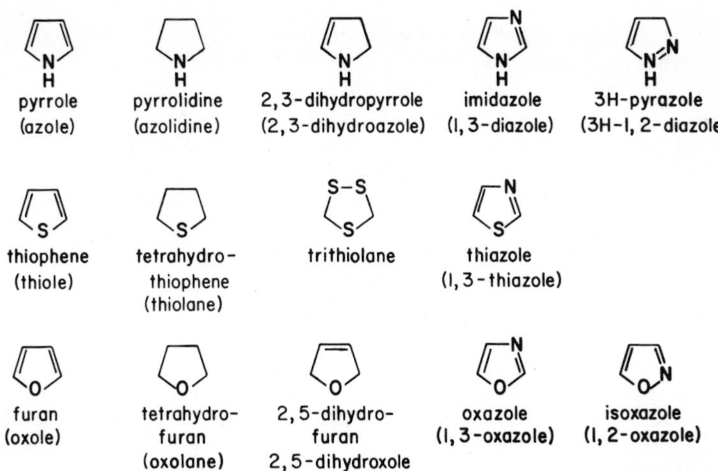

Figure 1. Five-membered ring heterocyclics

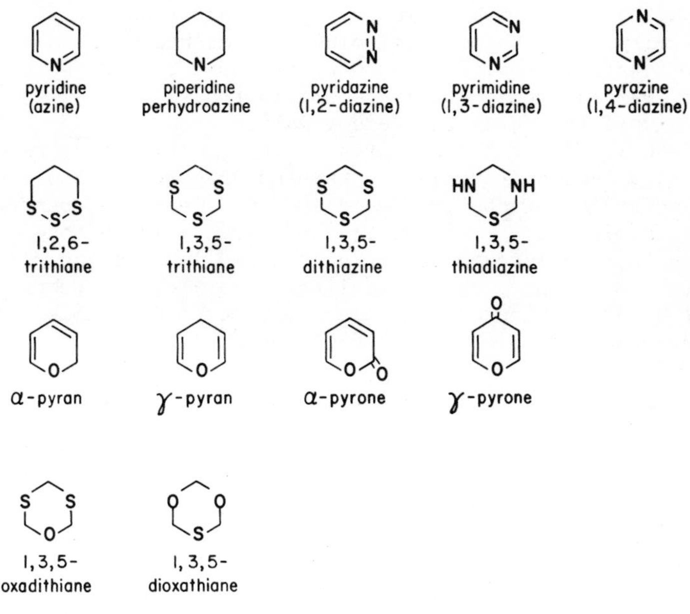

Figure 2. Six-membered ring heterocyclics

aldehyde were identified in the pyrrolic fraction of coffee volatile constituents (68, 69). N- ethylpyrrole-2-aldehyde and 5-methylpyrrole-2-aldehyde are two of the eighteen compounds reported in stored dehydrated orange juice crystals due to nonenzymic browning (70). The type of pyrrole derivatives produced depended on reaction temperature, its duration, its condition, i.e., anhydrous, aqueous, or alcoholic; pH of the media, and molecular structure of reactants. While, N-alkylpyrrole-2-aldehydes were produced upon heating D-xylose and alkylamine or amino acid in neutral aqueous or methanolic solutions at 55-100°C; the corresponding 5-methyl derivatives of N-substituted pyrrole-2-aldehydes were formed in the presence of L-rhamnose (71, 72). Odor of the products were: N-methylpyrrole-2-aldehyde (formed from D-xylose and methylamine) cinnamonaldehyde-like, N-n butylpyrrole-2-aldehyde (a product of butylamine-xylose); and 1-n-butyl-5-methylpyrrole-2-aldehyde (rhamnose and butylamine interactions product) xylene-like. 2-Formylpyrrol-1-yl-alkyl acids were isolated from the products of the browning reaction of xylose and alkylamines or amino acids in aqueous solutions (73). N-substituted-5-(hydroxymethyl)-pyrrole-2-aldehydes were formed from the reaction of the aldohexose "glucose" and alkylamines or amino acids containing primary amino group at 70 to 100°C, in neutralized aqueous, methanolic or ethanolic solutions. The resulting compounds were considerably unstable and had no odor in the pure state but developed roasted aroma with browning (74). Upon roasting glucose and several alkyl-n-amino acid (glycine, α-alanine, α-amino-n-butyric, valine, leucine, α-amino-n-caproic) at 200-250°C in two components systems; 2-5'-hydroxy-methyl-2'-formylpyrrol-1'-yl) alkanoic acid lactones were formed as the main volatile products (75). The aroma description of the prepared lactone derivatives were: propionic acid lactone (from α-alanine and glucose) caramel and a little scorching; isobutyric acid lactone (α-amino-n-butyric and glucose) maple and strong sweet; isovaleric acid lactone (valine and glucose) and isocarproic acid lactone (leucine and glucose) miso, soy sauce and a chocolate-like. The yield of propionic acid lactone after heating an equimolar mixture of 0.01 mole of glucose and α-alanine at 150, 200, and 250°C were:

Temperature, °C	Heating period, min.	Yield, unmoles
250	1	42
250	5	3 - 4
200	3	50
200	5	6
150	5	16

It is quite obvious from the above data that the propionic acid lactone disappeared during the reaction for longer periods and at higher temperatures. 2-(5'-hydroxymmethyl-2'-formylpyrrol-1'-yl) -3-methylbutanoate, and 2'-(5'-hydroxymethyl-2'-formylpyrrol-1'-yl)-3-methylbutanoic acid lactone were identified in the products of the browning reaction of glucose and valine in aqueous solution at 65°C for three weeks (76). Reducing disaccharides (lactose, maltose and melibiose) reacted with alkylamine in aqueous solutions of pH 6.5, to form 1-alkyl-5-hydroxymethylpyrrole-2-aldehyde (77). N-Alkyl-2-acylpyrroles and aliphatic aldimines were the products of the reaction between furfural and its homologs with α-amino acids (78). Reactions of furfural and 2-acetylfuran with glycine and valine produced a small amount of the corresponding acylalkylpyrroles which had pleasant aromas reminiscent of benzaldehyde. The resulting considerable amount of aliphatic aldimines from the reaction of furfural with valine and leucine possessed a strong odor from biting and unpleasant to mild and food-like. Nine alkylpyrroles, three acylpyrroles, three alkylpyrrole-2-aldehydes, three furfurylpyrrole, eight alkylpyrazines, and one oxazoline were identified in the volatile flavorous products of roasting DL-alanine and glucose at 250°C for one hour in nitrogen atmosphere (79). Many of the compounds identified in the reaction products of model systems had been isolated from roasted and cooked foods. For example, pyrrole, 1-methylpyrrole, 2-methylpyrrole, 1-formylpyrrole, 2-formylpyrrole, 1-formyl-2-methylpyrrole, 2-formyl-1-methylpyrrole, 2-formyl-1-methylpyrrole, 2-formyl-5-methylpyrrole, 2-formyl-1-ethylpyrrole, 1-acetylpyrrole, 2-acetylpyrrole, 1-furfurylpyrrole, 2-propionylpyrrole, 2-formyl-1-furfurylpyrrole were reported in roasted peanut (80). Roasted filberts volatiles contained 1-methylpyrrole, 2-pentylpyrrole, 2-isobutylpyrrole, and 2-penrylpyrridine (81).

Pyridine and its derivatives. The most unique pyridine derivative isolated from processed food is 1,4,5,6-tetrahydro-2-acetopyridine. This compound was prepared by roasting proline and dihydroxyacetone at 92°C in presence of sodium bisulfate, and exhibited a strong odor reminiscent of freshly backed soda crackers (82). 2-Ethylpyridine and 2-pentylpyridine were reported in volatile flavor components of shallow fried (83). Pyridine, 2-methylpyridine, 3-methylpyridine, 2-ethylpyridine, 3-ethylpyridine, 5-ethyl-2-methylpyridine, 2-butylpyridine, 2-acetylpyridine, 2-pentylpyridine, 2-hexylpyridine, 3-pentylpyridine, 5-methyl-2-pentylpyridine, and 5-ethyl-2-pentylpyridine were identified in the volatiles of roasted lamb fat (84). 2,5-Dimethylpyridine and 3,5-dimethylpyridine were tentatively identified in roasted lamb fat volatiles. The odor threshold of 2-pentylpyridine was 0.5-0.7 parts per 10^9 parts of water. The dilute solution of 2-pentylpyridine has a fatty or tallow-like odor. The authors attribute the unacceptance of lamb by some consumers to the high content of alkylpyridines in roasted lamb. While pyridine derivatives have burnt, heavy fruity odors (85); pyridine has a disagreeable characterisitic odor and sharp taste; and piperidine has a peppery ammoniacal odor (86). Pyridine, α-picoline (2-methylpyridine), β-picoline (4-methylpyridine), and 4-ethylpyridine were identified in coffee aroma (87).

Pyrazines. In the thirties, the attention on pyrazines was focused on its industrial role in dyes, photographic emulsions and chemotherapy. Its importance in life processes was indicated in its derivative, vitamin B_2 (riboflavin, 6,7-dimethyl-9-(1'-D-ribityl isoalloxazine). Later,in the midsixties, it was identified in foods and its contributions to the unique flavor and aroma of raw and processed foods attracted the attention of flavor chemists. Pyrazine derivatives contribute to the roasting, toasting, nutty, chocolaty, coffee, earthy, caramel, maple-like, bread-like, and bell pepper notes in foods. The reader is referred to the reviews on Krems and Spoerri (88) on the chemistry of pyrazines, and the review of pyrazines in foods by Maga and Sizer (89, 90). Table XVI summaries sensory description and threshold of selected pyrazines.

Oxygen Heterocyclics. During heat processing, sugars degrade to aldehydes and ketones which might react with amino compounds forming caramelized sugar flavors. Cyclic diketones, pyrones, and furan derivatives are examples of the products of this reaction. Table XVII gives the organoleptic description of selected compounds. 4-Hydroxy-2,5-dimethyl(2H)-furanone has intense fragrant caramel note described as burnt sweet taste (70), burnt pineapple (101), beef broth (100), strawberry preserve (103), nutty sweet aroma of almonds (104), and major contributor to sponge cake flavor (105). Seven terms were used to describe the organoleptic note of one compound. The question arises, what was the concentration and media used for evaluation? Change in concentration alone can be quite sufficient to alter the character of the flavor or odor note. For example, trimethylamine-air mixtures only smell fishy over a narrow range of dilutions (1:1,500-1:8,000), with a maximum at about (1:6,000), also ammonia at a dilution of about (1:2,000) smells fishy (106). Hodge and Moser (107) reported that inspite of the fact that the aromas of pure maltol, ethyl maltol, isomaltol, and 4-hydroxy-2,5-dimethyl-3(2H)-furanone vary significantly from each other, panelists description was caramel or burnt sugar. The fruity caramel aroma of isomaltol is weaker than that of 4-hydroxy-2,5-dimethyl-3(2H)-furanone (108). Our sensory vocabulary should be adequate to express the impact of flavor, odor, taste notes components. This could be achieved by the participation of food chemist, organic chemist, sensory analyst and flavorist. Soy sauce (Shoyu) contains tautomers 4-hydroxy-2-ethyl-5-ethyl-3(2H)-furanone and 4-hydroxy-5-ethyl-2-methyl-3(2H)-furanone (about 3:2 ratio) which has sweet odor similar to that of short cake (109). The same tautomers were synthesized and described as possessing the flavor of cooked fruits (103). Maltol, isomaltol and 2-methyl-5-hydroxy-6-ethyl-α-pyrone are contributors to the characteristic aroma of molasses (110).

Table XVI

Pyrazines, Sensory Description and Odor Threshold

Pyrazine	Sensory Description	Odor Threshold ppm Water	Mineral Oil	Vegetable Oil
2-Methyl-	deep bitter persistent cocoa notes (5)	105 (91) 60 (92)	27 (91)	
2,3-Dimethyl-		2.5 (92)	--	
2,5-Dimethyl-	earthy raw potato (3)	3.5 (91) 1.8 (92)	7 (91)	2.6 (91)
2,6-Dimethyl	sweet fried odor (2)	54 (91) 1.5 (92)	8 (91)	
2,3,5-Tri-methyl		9 (91)	27 (91)	
2,3,5,6-Tetra-methyl		10 (91)	38 (91)	
2-Ethyl		60 (92) 22 (91)	17 (91)	
2-Ethyl-3-methyl-		0.13 (92)		0.32 (92)
2-Ethyl-5-methyl		0.10 (92)		
2-Ethyl-3,6-dimethyl-		0.0004 (92)	--	0.024 (92)

Table XVI

Pyrazines, Sensory Description and Odor Threshold (Cont.)

Pyrazine	Sensory Description	Odor Threshold ppm Water	Mineral Oil	Vegetable Oil
2,6-dimethoxy-3-isopropyl-5-methyl	nutty with green-like bell-pepper, woody notes (96)			
2,5-dimethoxy-3-isopropyl-6-methyl	nutty notes, green (bell-pepper-like) (96)			
5-methyl 5H-6,7-dihydrocyclopenta	peanuts (53)			
5-methyl-2H-3,4,5,7-retrahydrocyclopenta	roasted nuts, burnt (53)			
2-acetyl-	breadcrust, nutty (96)			
2-methoxy-3-acetyl-	weak breadcrust, nutty notes (96)			
2-methylamino-3-methyl	roasted peanuts, green cocoa notes (96)			
2-dimethylamino-3-methyl	Peanut-like more green and less cocoa notes (96)			
2-dimethylamino-6-methyl	peanut-like, no cocoa notes reminiscent of burnt coffee (96)			

Table XVI

Pyrasines, Sensory Description and Odor Threshold (Cont.)

Pyrazine	Sensory Description	Water	Mineral Oil	Vegetable Oil
2,5-diethyl-		0.02 (92)		0.27 (92)
2,6-diethyl-		0.006 (92)		
5-Ethyl-2,3-dimethyl	chocolate sweet (93)			
2,5-Dimethyl-3-ethyl		43 (91)	24 (91)	
2,6-Dimethyl-3-ethyl		15 (91)	24 (91)	
Methyl isoamyl-	green cocoa note (95)			
Ethyl isoamyl-	green cocoa note (95)			
2-Methylamine-3-methyl-	cocoa notes, roasted peanuts, green (96)			
2-Dimethylamine-3-ethyl	less cocoa note (96)			
2-Methyl-3-methoxy-	roasted peanut (91) nutty earthy (97)			
2-Methylthio-3-methyl-	nutty cracker (97)			

References (53, 91-97).

Table XVII

Oxygen Heterocyclic Compounds

Compound	Sensory Description
Furan	Spicy, smoky, slightly cinnamon-like odor (98).
	Sweet bread-like, caramel-like taste (98).
Furfuryl alcohol	Coconut (99).
5-methyl-2-furfural	Sharp grape (98).
Maltol (3-hydroxy-2-methyl-pyrone)	Pleasing strong fruitsy fresh bread (99).
Isomaltol	Fruity caramel, fresh bread odor, overtone medicinal grassy (99).
N-Furfuryl pyrrole	Green hay-like aroma (98).
4-hydroxy-5-methyl-3(2H)-Furanone	Beef broth (100) burning sweet taste (70), burnt pineapple (101).
2,5-Dimethyl-4-hydroxy-3(2H)-Furanone	Odor of roasted chicory roots (102).

References (98 - 102).

Sulfur Heterocyclics. Sulfur containing compounds (thiols, thiophenes, thiazoles, ... etc.) play a major role in the flavor of raw and processed foods. These compounds have characteristic flavor notes and the flavor thresholds are mostly low. Several reviews (111, 112, 113) demonstrate the important role of sulfur compounds in food flavors. Organoleptic properties of these compounds may be pleasant, strong nut-like odor of 4-methyl-5-vinylthiazole which is present in cocoa (114); objectionable pyridine-like odor of thiazole (115); quinoline-like odor of benzothiazole (116); strong tomato leaf-like odor of isobutylthiazole (117); and bread crust flavor of acetyl-2-thiazoline (118). A mixture of oxazoles, thiazoles, thiazolines, imidazoles, trithiolanes and dithianes which had a meaty flavor was obtained from a model system consisting of α-dicarbonyls, ethanal, hydrogen sulfide and ammonia (119). Unsaturated aldehydes react with hydrogen sulfide and thiols to give mainly addition products to the carbon-carbon double bond (120). The nomenclature of the resulting compounds and their organoleptic descriptions are given in Table XVIII. Thiols, thiophenes, thiazoles, sulfides and furans were identified in the volatiles of heating glucose, hydrogen sulfide, and ammonia at 100°C for two hours (121). These volatiles gave a roast beef-like aroma. Complex mixtures of mercapto-substituted furans and thiophene derivatives, which were reminiscent of roast meat were produced upon heating 4-hydroxy-5 methyl-3(2H)-furanone and its thio analog with hydrogen sulfide (122).

Lipid Browning.

Lipid browning reactions of the Maillard type between carbonyl groups (provided by sugars or sugar degradation products and those resulting from unsaturated fatty acids oxidation) and the free amino groups present in phospholipids have been recognized as potent causes of undesirable flavor, color and texture changes in dehydrated foods (123). Phosphatidyl ethanolamine, phoshatidyl serine, ethanolamine and serine plasmalogens contain free amino groups which can undergo lipid browning reactions. Phospolipids are usually rich in highly unsaturated fatty acids in comparison with neutral lipids, thus they are good sources of carbonyls. Also, the primary amine moities of polar lipids catalyze the aldol condensation of C_{14}-C_{18} aldehydes resulting from plasmalogen hydrolysis, thus forming α,β-unsaturated aldehydes (124). Phosphatidyl ethanolamine reacted with propanal and n-hexanal forming phosphatidyl 1-(2-hydroxyethyl)-2-ethyl-3,5-dimethyl pyridinium, and phosphatidyl-1-(2-hydroxyethyl)-2-hexyl-3,5-dipentyl pyridinium, respectively (125). The peridinium ring is formed by the reaction between one mole of amino-N of phosphatidyl ethanolamine and three moles of n-alkanals. The same reaction took place in the synthesis of substituted pyridines by condensation of carbonyl compounds with ammonia (126, 127).

Abstract

In processed foods, non-enzymic browning reaction is the major source of its desirable flavors. Flavors of the products of this reaction depend upon: the molecular structure of nitrogenous compounds (amines, amino acids, peptides, glycopeptides, proteins, ... etc.); aldoses, ketoses, non-reducing, deoxy sugars, sugar acids, ... etc.); heating temperature; duration of the reaction; initial hydrogen ion concentration, moisture content, and the media of the reaction (alcoholic, aqueous, or anhydrous). The ratio of the nitrogenous compound to sugar or carbonyl compound has great effect on the flavor notes. Comparison betwen browning

Table XVIII

Flavor of Products of the Addition Between Unsaturated Carbonyls and Methanethiol

Carbonyl Added	Addition Product	Flavor Impression	Water	Paraffin Oil
2-Butenal	3-methyl thiobutanal	Cheese-like (Brite)	0.5	0.5
2-Hexenal	3-Methyl thiohexanal	Cabbage (Rubbery)	5	50
2-Hebtenal	3-Methyl thioheptanal	Unripe tomato	5	80
2-Nonenal	3-Methyl thiononal	Bast-like, slightly floral	3	20
1-Octene-3-one	1-Methyl thiooctan-3-one	Radish-like		

Reference (120).

reaction products and rate of degradation of reactants under anhydrous condition, aqueous and alcoholic media was discussed. Flavor notes of products of amino compounds-sugars reaction were reviewed with emphasis on amino-ribose system. Meat-like flavor was imparted by browning reaction products of carnosine-, citrulline-, histidine-, glutamic-, 2-pyrrolidone-5-carboxylic-, methionine-, cysteine-, cysteic-, and taurine-ribose. Recent advancements in nitrogen-, oxygen- and sulfur hetercyclics and lipid browning were presented.

Literature Cited.

1. Economics, Statistics, and Cooperative Service, USDA, National Review, 1978, April; p. 53.
2. Yamasaki, Y.; Kaekawa, K., Agr. Biol. Chem., 1978, 42, 1761.
3. Waley, S.G., Advan. Protein Chem., 1966, 21, 1.
4. Reineccius, G.A.; Andersen, D.A.; Kavanagh, T.E.; Keeney, P.G., J. Agr. Food Chem., 1972, 20, 1.
5. Patterson, R.L.S., J. Sc. Food Agr., 1968, 19, 31.
6. Wong, E.; Nixon, L.N.; Johnson, C.B., J. Agr. Food Chem., 1975, 23, 495.
7. Watanabe, K.; Sato, Y., Agr. Biol. Chem., 1970, 34, 88.
8. Watanabe, K.; Sato, Y., Agr. Biol. Chem., 1971, 35, 756.
9. Yamato, T.; Tadao, K.; Kato, H.; Fujimaki, M., Agr. Biol. Chem., 1970, 34, 88.
10. Hornstein, J., Chemistry and Physiology of Flavors, Schultz, H.W.; Day, E.A.; Libbey, L.M., Eds., Avi Publishing Co., Westport, CT, 1967; pp. 228-250.
11. Watanabe, K.; Sato, Y., Agr. Biol. Chem., 1970, 34, 467.
12. Watanabe, K.; Sato, Y., Agr. Biol. Chem., 1971, 35, 278.
13. Harkes, P.D.; Begemann, W.J., J. Am. Oil Chemists' Soc., 1974, 51, 356.
14. Hornstein, I.; Crowe, P.F.; Heimberg, M.J., J. Food Sci., 1961, 26, 581.
15. Pippen, E.L.; Nonaka, M.; Jones, F.T.; Stitt, F., Food Res., 1958, 23, 103.
16. Pippen, E.L.; Nonaka, M., Food Res., 1960, 25, 764.
17. Heyns, K.; Stute, R.; Paulsen, H., Carbohydrate Res., 1966, 2, 132.
18. Heyns, K.; Klier, M., Carbohydrate Res., 1968, 6, 436.
19. Walter, R.; Fagerson, I.S., J. Food Sci., 1968, 33, 294.
20. Kort, M.J., Advan. Carbohydrate Chem. Biochem., 1970, 25, 311.
21. Chuyen, N.V.; Kurata, T.; Fujimaki, M., Agr. Biol. Chem., 1973, 37, 327.
22. Kato, S.; Kurata, T.; Fujimaki, M., Agr. Biol. Chem., 1973, 37, 1759.
23. Fujimaki, M.; Kato, S.; Kurata, T., Agr. Biol. Chem., 1969, 33, 1144.
24. Kato, S.; Kurata, T.; Fujimaki, M., Agr. Biol. Chem., 1973, 37, 539.

25. Schonberg, A.; Moubacher, R., Chem. Rev., 1952, 50, 261.
26. Hodge, J.E., J. Agr. Food Chem., 1953, 1, 928.
27. Hodge, J.E., Advan. Carbohydrate Chem., 1955, 10, 169.
28. Ellis, G.P., Advan. Carbohydrate Chem., 1959, 14, 163.
29. Reynolds, T.M., Advan. Food Res., 1963, 12, 1.
30. Reynolds, T.M., Advan. Food Res., 1965, 14, 167.
31. Lightbody, H.D.; Feuold, H.R., Advan. Food Res., 1948, 1, 149.
32. Ross, A.F., Advan. Food Res., 1948, 1, 325.
33. Stadman, E.R., Advan. Food Res., 1948, 1, pp. 325-372.
34. Wodemeyer, G.A.; Dollar, A.M., J. Food Sci., 1963, 28, 537.
35. Wiseblatt, L.; Zoumut, H.F., Cereal Chem., 1963, 40, 162.
36. El'Ode, K.E.; Dornseifer, T.P.; Keith, E.S.; Powers, J.J., J. Food Sci., 1966, 31, 351.
37. Herz, W.J.; Shallenberger, R.S., Food Res., 1960, 25, 491.
38. Barnes, H.M.; Kaufman, C.W., Ind. Eng. Chem., 1947, 39, 1167.
39. Kiley, P.J.; Nowlin, A.C.; Mariarty, J.H., Cereal Sci. Today, 1960, 5, 273.
40. Buckdeschel, W.Z., Ges Brau, 1914, 430, thru Chem. Abs., 1915, 9, 118.
41. Morimoto, T.; Johnson, J.A., Cereal Chem., 1966, 43, 627.
42. Keeney, M.; Day, E.A., J. Dairy Sci., 1957, 40, 847.
43. Actander, S., Perfume and Flavor Chemicals (Aroma Chemicals); Published by the Author, Montclair, NJ, 1969; two volumes.
44. Persson, T.; von Sydow, E., J. Food Sci., 1973, 38, 377.
45. Rooney, L.W.; Salem, A.; Johnson, J.A., Cereal Chem., 1967, 44, 539.
46. Self, R., Chemistry and Physiology of Flavors, Schultz, H.W.; Day, E.A.; Libbey, L.M., Eds., Avi Publishing Co., Westport, CT, 1967; pp. 362-389.
47. Mabrouk, A.F.; Holms, L.G., Abstracts of Papers, 159, Division of Agricultural and Food Chemistry, 172nd ACS National Meeting, San Francisco, CA, September 1976.
48. Rohan, T.A.; Stewart, T., J. Food Sci., 1966, 31, 202.
49. Rohan, T.A.; Stewart, T., J. Food Sci., 1966, 31, 206.
50. Rohan, T.A.; Stewart, T., J. Food Sci., 1967, 32, 395.
51. Rohan, T.A.; Stewart, T., J. Food Sci., 1967, 32, 399.
52. Rohan, T.A.; Stewart, T., J. Food Sci., 1967, 32, 625.
53. van Praag, M.; Stein, S.H.; Tibbets, M.S., J. Agr. Food Chem., 1968, 16, 1005.
54. Pickenhagen, W.; Dietrich, P.; Keil, B.; Polansky, J.; Nouaille, F.; Lederer, E., Helv. Chem. Acta., 1975, 58, 1078.
55. Kato, S.; Kurata, T.; Ishitsuka, R.; Fujimaki, M., Agr. Biol. Chem., 1970, 34, 1826.
56. Shiba, T.; Nunami, K.I., Tetrahedron Letters, 1974, 6, 509.
57. Takanishi, K.; Tadenuma, M.; Kitamoto, K.; Sato, S., Agr. Biol. Chem., 1974, 38, 927.
58. Sakamura, S.; Furukawa, K.; Kasai, T., Agri. Biol. Chem., 1978, 42, 607.
59. Boyd, E.N.; Keeney, P.G.; Patton, S., J. Food Sci., 1965, 30, 854.

60. van Straten, S.; de Vrijer, F., List of Volatile Compounds in Food, Central Institute for Nutrition and Food Research, Zeist, Holland, 1973.
61. Eckey, E.W., Vegetable Fats and Oils, Reinhold Publishing Corp., NY, 1954; p. 676, 760.
62. Feldman, J.R.; Ryder, W.S.; Kung, J.T., J. Agr. Food Chem., 1969, 17, 733.
63. Underwood, G.E.; Deatherage, F.E., Food Res., 1952, 17, 425.
64. Khan, N.A.; Brown, J.B., J. Am. Oil Chemists' Soc., 1953, 30 606.
65. Shigematsu, H.; Kurata, T.; Kato, H.; Fujimaki, M., Agr. Biol. Chem., 1972, 36, 1631.
66. Kobayasi, N.; Fujimaki, M., Agr. Biol. Chem., 1965, 29, 1059.
67. Yoshikawa, K.; Libbey, L.M.; Cobb, W.Y.; Day, E.A., J. Food Sci., 1965, 30, 991.
68. Gianturco, M.A.; Giammarino, A.S.; Friedel, P.; Flanagan, V., Tetrahedron, 1964, 20, 2951.
69. Gianturco, M.A.; Giammarino, A.S.; Friedel, P., Nature, 1966, 210, 1358.
70. Tatum, J.H.; Shaw, P.E.; Berry, R.E., J. Agr. Food Chem., 1967, 15, 773.
71. Kato, H., Agr. Biol. Chem., 1966, 30, 822.
72. Kato, H., Agr. Biol. Chem., 1967, 31, 1086.
73. Kato, H.; Fujimaki, M., J. Food Sci., 1968, 33, 445.
74. Kato, H.; Fujimaki, M., Agr. Biol. Chem., 1970, 34, 1071.
75. Shigematsu, H.; Kurata, T.; Kato, H.; Fujimaki, M., Agr. Biol. Chem., 1971, 35, 2097.
76. Kato, H.; Sonobe, H.; Fujimaki, M., Agr. Biol. Chem., 1977, 41, 711.
77. Sonobe, H.; Kato, H.; Fujimaki, M., Agr. Biol. Chem., 1977, 41, 609.
78. Rizzi, G.P., J. Agr. Food Chem., 1974, 22, 279.
79. Shigematsu, H.; Kurata, T.; Kato, H.; Fujimaki, M., Agr. Biol. Chem., 1972, 36, 1631.
80. Walradt, J.P.; Pittet, A.O.; Kinlin, T.E.; Muralidhara, R.; Sanderson, A., J. Agr. Food Chem., 1971, 19, 972.
81. Kinlin, T.E.; Muralidhara, R.; Pittet, A.O.; Sanderson, A.; Walradt, T.P., J. Agr. Food Chem., 1972, 20, 1021.
82. Hunter, I.R.; Walden, M.K.; Scherer, J.R.; Lundin, R.E., Cereal Chem., 1969, 46, 189.
83. Watanabe, K.; Sato, Y., J. Agr. Food Chem., 1971, 19, 1017.
84. Buttery, R.G.; Ling, L.C.; Teranishi, R.; Mont, T.R., J. Agr. Food Chem., 1977, 25, 1277.
85. MacLeod, G.; Coppock, B.M., J. Agr. Food Chem., 1977, 25, 113.
86. Webster's Third New International Dictionary, G&C Merriam Co., Publishers, Springfield, MA, 1967.
87. Goldman, I.M.; Seibl, J.; Flament, I.; Gautschi, F.; Winter, M.; Willham, B.; Stoll, M., Helv. Chem. Acta., 1967, 50, 694.
88. Krems, I.J.; Spoerri, P.E., Chem. Rev., 1947, 40, 279.

89. Maga, J.A.; Sizer, C.E., Crit. Rev. Food Technol., 1973, 4, 39.
90. Maga, J.A.; Sizer. C.E., J. Agr. Food Chem., 1973, 21, 22.
91. Seifert, R.M.; Buttery, R.G.; Guadagni, D.G.; Black, D.R.; Harris, J.G., J. Agr. Food Chem., 1970, 18, 246.
92. Seifert, R.M.; Buttery, R.G.; Guadagni, D.G.; Black, D.R.; Harris, J.G., J. Agr. Food Chem., 1972, 20, 135.
93. Deck, R.E.; Chang, S.S., Chem. Ind., 1965, 1343.
94. Buttery, R.G.; Seifert, R.M.; Guadagni, D.G.; Ling, L.C., J. Agr. Food Chem., 1969, 17, 1322.
95. Polak's Fruital Works, Improvements relating to flavorings, Br. Patent 1,248,380, September 29, 1971.
96. Takken, H.J.; van der Linde, L.M.; Boelens, M.; van Dort, J. M., J. Agr. Food Chem., 1975, 23, 638.
97. Parliment, T.H.; Epstein, M., J. Agr. Food Chem., 1973, 21, 714.
98. Guadagni, D.G.; Buttery, R.G., J. Sci. Food Agr., 1972, 23, 1435.
99. Shaw, P.E.; Tatum, J.H.; Berry, R.E., J. Agr. Food Chem., 1969, 17, 907.
100. Tonsbeek, C.H.T.; Planchen, A.J.; van de Weerdhof, T., J. Agr. Food Chem., 1968, 16, 1016.
101. Rodin, J.O.; Himel, C.M.; Silverstein, R.M.; Leeper, R.W.; Gortner, W.A., J. Food Sci., 1965, 30, 280.
102. Tonsbeek, C.H.T.; Koenders, E.B.; van der Zijden, A.S.M.; Losekoot, J.A., J. Agr. Food Chem., 1969, 17, 397.
103. Re, V.L.; Maurer, B.; Ohloff, G., Helv. Chem. Acta., 1973, 56, 1882.
104. Takei, Y.; Yamanishi, T., Agr. Biol. Chem., 1974, 38, 2329.
105. Takei, Y., Agr. Biol. Chem., 1977, 41, 2361.
106. Stansby, M.E., Food Technol., 1962, 16, 28.
107. Hodge, J.E.; Moser, H.A., Cereal Chem., 1961, 38, 221.
108. Hodge, J.E.; Mills, F.D.; Fisher, B.E., Cereal Sci. Today, 1972, 17, 34.
109. Numomura, N.; Saaki, M.; Asao, Y.; Yokotsuka, T., Agr. Biol. Chem., 1976, 40, 491.
110. Ito, H., Agr. Biol. Chem., 1976, 40, 827.
111. Schutte, L., Crit. Rev. Food Technol., 1974, 4, 457.
112. Maga, J.A., Crit. Rev. Food Sci. Nutr., 1975, 6, 153.
113. Maga, J.A., Crit. Rev. Food Sci. Nutr., 1975, 6, 241.
114. Maga, J.A., Crit. Rev. Food Sci. Nutr., 1976, 7, 147.
115. Stoll, M.; Dietrich, P.; Sunte, E.; Winter, M., Helv. Chem. Acta., 1967, 50, 2065.
116. Merck Index 9th ed., 1976; p. 1313.
117. Viani, R.; Bricout, J.; Marion, J.P.; Muggler-Chavan, F.; Reymond, D.; Egli, R.H., Helv. Chem. Acta., 1969, 52, 887.
118. Tonsbeek, C.H.T.; Copier, H.; Plancken, A.J., J. Agr. Food Chem., 1971, 19, 1014.

119. Boelens, M.; van der Linde, L.M.; de Valois, P.J.; van Dort, H.M.; Takken, H.J., Proc. Int. Symp. Aroma Research, Zeist, Netherland, 1975; pp. 95-100.
120. Badings, H.T.; Maarse, H.; Kleipool, R.J.C.; Tas, A.C.; Neeter, R.; ten Noever de Brauw, M.C., Proc. Int. Symp. Aroma Research, Zeist, Netherland, 1975; pp. 63-73.
121. Shibamoto, T.; Russell, G.F., J. Agr. Food Chem., 1976, 24, 843.
122. van den Ouweland, G.A.M.; Peer, H.G., J. Agr. Food Chem., 1975, 23, 501.
123. Lea, C.H., J. Sci. Food Agr, 1957, 8, 1.
124. Nakanishi, T.; Suyama, K., Nippon Chikusan Gakkai-Ho, 1970, 41, 29.
125. Nakanishi, T.; Suyama, K., Agr. Biol. Chem., 1974, 38, 1141.
126. Frank, R.L.; Seven, R.P., J. Amer. Chem. Soc., 1949, 71, 2629.
127. Farley, C.P.; Eliel, E.L., J. Amer. Chem. Soc., 1956, 78, 3477.

RECEIVED August 2, 1979.

10

Concluding Remarks

MITSUO NAMIKI

Department of Food Science and Technology, Nagoya University, Chikusa-ku, Nagoya, Japan

Dear friends, we are very happy to have completed this really unique Symposium with such a success. As one of the organizers of this symposium, I would like to say a few words to close this session on taste. Since this symposium is held in the Joint Chemical Congress of ACS and CSJ, I hope you will allow me to talk a little while about my rather mixed up ideas on taste by using Japanese language.

Figure I shows how we write "Symposium on the Taste of Foods" in Japanese. The first example is simply the phonetic transliteration of the word Symposium written in "Katakana", a kind of Japanese traditional script. The next two Chinese characters are pronounced "shokuhin" and mean "food", the last character is "aji" which means "taste" or "flavor", our common interest. The character between them is "no" in "Hiragana", another kind of Japanese traditional script, and means "of".

As you see, the Japanese language uses three kinds of scripts with the words completely reversed from the arrangement found in European languages. So I hope you might have some sympathetic feelings for our Japanese scientists here, including myself.

Anyway, let me tell you how I did my little analytical work on these words. Since Chinese is picture writing, the character for "aji" can be separated into two components. The left one is simply a square, meaning mouth. The other half is considered to be phonetically equivalent to "mi" or "bi" which means "beauty" or "goodness". Therefore the composition of the character for "aji" indicates that taste is primarily "good to mouth", namely, palatable, delicious, and tasty. In this sense, I heartily agree with Dr. Boudreau who wisely pointed out that "umami" should be counted as a basic taste.

The Chinese character for "umami", as shown in this figure, has its origin in "a spoon and mouth", namely a good taste elicited by eating a delicious soup. As Dr. Yamaguchi stated, this taste "umami" is known to be attributed to mainly two factors, MSG and nucleotides. It happens that the use of both of these factors as flavoring agents was originated by Japanese scientists ;

0-8412-0526-4/79/47-115-247$05.00/0

Symposium The Taste of Foods

シンポジウム 食品の味

あじ	あまい	にがい	すっぱい	しおからい	うまい	からい
味	甘	苦	酸	塩	旨	辛
Taste	Sweet	Bitter	Sour	Salty	Umami	Pungent

味 = 口 + 未

Taste Flavor — Mouth — 美 (mi) (bi) Beauty Good Grace

旨 < spoon mouth

Figure 1.

MSG by Dr. Kikunae Ikeda, who was once the president of the Chemical Society of Japan, and nucleotides by Dr. Kodama, Dr. Kuninaka and others. Dr. Kuninaka's finding that a small amount of nucleotide enhanced remarkably the "umami" of MSG has been of interest to the world, especially to the food industry. Here it is interesting to note that in the Japanese language the word "umami" sometimes means "unexpected profit" which just fits this case. I might add that manufacture of these flavoring agents has developed into a big fermentation industry, which was developed mainly by the efforts of the agricultural chemists of Japan.

The umami taste, however, is not always caused by just these two agents, rather it is elicited by a well harmonized mixture of various food constituents, such as amino acids, peptides, nucleotides, organic acids, and inorganic ions. This fact is clearly demonstrated by the studies of Dr. Konosu on the typical umami of seafoods, and by Dr. Mori on the taste of soy sauce.

Speaking of the mixture of these flavoring agents, I think Japanese people have a good background to develop "umami". As you have seen in the Japanese letters, the Japanese people are themselves a well homogenized mixture both in cultural and ethnic aspects. Their food habits are typically omnivorous. Moreover, owing to the varying climate and a marine island country, the food is abundant in kind, varying from raw fish to various delicious fermentation products as shoyu, miso and, especially, sake.

Here, I must call attention to the reports by Dr. Solms and Dr. Mabrouk who informed us that the umami substances were developed and increased by cooking and processing of foods. It is these types of studies combined with a sophisticated psycophysics which will help us better understand food flavor.

Now, I feel I have dwelt too long on the subject of "umami", but the word "umami" in Japanese language sometimes means "sweetness". As to the sweeteners, it is no wonder that such a great deal of work has been done on new sweeteners of natural and artificial origin. Until now, such work has been a kind of hit and miss business. Therefore, the last half of the day was devoted to understanding some of the structural features of molecules that determine their taste properties. Based on the advanced stereo-chemical studies on a large number of sweet and bitter compounds by Dr. Ariyoshi, Dr. Belitz and Dr. Ney, our understanding of the molecular properties of certain taste compounds has advanced markedly.

Noticeable also is an increase in our understanding of the chemical properties of amino acids, peptides and similar nitrogen compounds, since as we saw in the first half of the symposium, they are primary flavor components in many foods.

At this point, I would like to indicate that taste chemistry is essentially solution chemistry. It is therefore especially significant that the bitterness of many compounds can be directly related to the hydrophobic properties of the molecules. The sour sensation has also been shown to be related to the Brønsted acid properties of the molecules. Thus, both sour and bitter tastes can be shown to be related to the solution properties of the molecules. In the final analysis, that which stimulates is a molecular complex of the active molecule and water. Solution chemistry is in as primitive a state as is taste chemistry and they must develop together.

On another important taste, pungency, Dr. Govindarajan said in his paper that this sensation must be measured to accurately describe the flavor of certain foods. He has also suggested a relationship between taste and chemical structure as has already been done in the cases of other tastes.

Largely untouched by this symposium, but possibly present in the minds of many, is the nutritional significance of taste. What is the exact relationship between taste and nutrition? If we cook foods to produce flavorful compounds, what is the nutritional significance of these compounds? These and many other questions await answers.

In ending these final remarks, I must state that I feel so happy to learn directly that the efforts of scientists all over the world are resulting in a steady progress in this difficult field, and are contributing to the welfare of men by appealing to their most fundamental and peaceful desires of eating good things. I believe this occasion will be an important milestone in the development of this research.

Finally, I would like to extend my hearty thanks to the speakers for performing so well, the audience for being so attentive and to Dr. Boudreau for his great efforts in organizing this symposium.

Thank you.

RECEIVED August 16, 1979.

INDEX

INDEX

M

N

O

P

Q

R

S